Advanced Maths Essentials Core 4 for Edexcel

Welcome to Advanced Maths Essentials: Core 4 for Edexcel. This book will help you to improve your examination performance by focusing on essential maths skills you will need in your Edexcel Core 4 examination. It has been divided by chapter into the main topics that need to be studied. Each chapter has then been divided by sub-headings, and the description below each sub-heading gives the Edexcel specification for that aspect of the topic.

The book contains scores of worked examples, each with clearly set-out steps to help solve the problem. You can then apply the steps to solve the Skills Check questions in the book and past exam questions at the end of each chapter. If you feel you need extra practice on any topic, you can try the Skills Check Extra exercises on the accompanying CD-ROM. At the back of this book there is a sample exam-style paper to help you test yourself before the big day.

Some of the questions in the book have a symbol next to them. These questions have a PowerPoint® solution (on the CD-ROM) that guides you through suggested steps in solving the problem and setting out your answer clearly.

Using the CD-ROM

To use the accompanying CD-ROM simply put the disc in your CD-ROM drive, and the menu should appear automatically. If it doesn't automatically run on your PC:

1. Select the My Computer icon on your desktop.
2. Select the CD-ROM drive icon.
3. Select Open.
4. Select core4_for _edexcel.exe.

If you don't have PowerPoint® on your computer you can download PowerPoint 2003 Viewer®. This will allow you to view and print the presentations. Download the viewer from http://www.microsoft.com

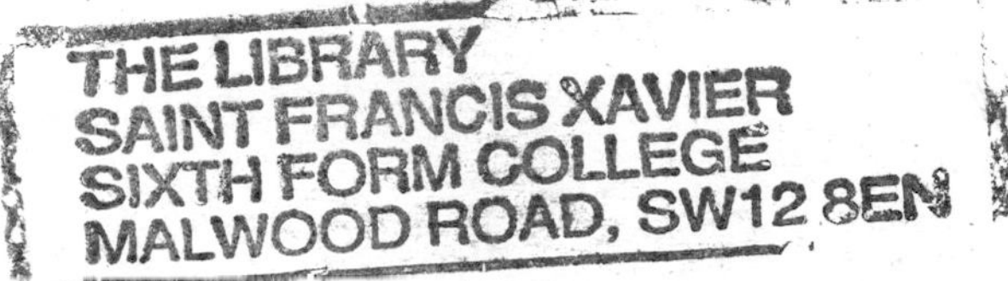

Pearson Education Limited
Edinburgh Gate
Harlow
Essex
CM20 2JE
England
www.longman.co.uk

First published 2006
ISBN-13: 978-0-582-83672-3
ISBN-10: 0-582-83672-7

Design by Ken Vail Graphic Design

Cover design by Raven Design

Typeset by Tech-Set, Gateshead

Printed in the U.K. by CPI Bath Press

The publisher's policy is to use paper manufactured from sustainable forests.

We are grateful for permission from the London Qualifications Limited trading as Edexcel to reproduce past exam questions. All such questions have a reference in the margin. London Qualifications Limited trading as Edexcel can accept no responsibility whatsoever for accuracy of any solutions or answers to these questions.

1 Algebra and functions

1.1 Partial fractions

Rational functions. Partial fractions.

In *Core 3* you added and subtracted algebraic fractions.

For example, you showed that

$$\frac{5}{x-1}+\frac{3}{x+2}=\frac{8x+7}{(x-1)(x+2)}$$

In the reverse of this process, a rational function is expressed as a sum of two or more simpler fractions called **partial fractions**.

You will need to be able to split the following proper and improper fractions into partial fractions:

- fractions with distinct linear factors in the denominator
- fractions with a repeated linear factor in the denominator.

Recall: Algebraic fractions (C3 Section 1.1).

Note: The method of splitting a single fraction into partial fractions is also needed in C4 for work on the binomial series and integration.

Fractions with distinct linear factors in the denominator

Notice that in the above example

$$\frac{8x+7}{(x-1)(x+2)}=\frac{5}{x-1}+\frac{3}{x+2}$$

the denominators of the partial fractions are the factors of the original denominator. When splitting fractions of this type into partial fractions use this fact as a starting point.

Example 1.1 Express $\frac{2x-4}{x^2-2x-3}$ in partial fractions.

Step 1: Factorise the denominator.

$$\frac{2x-4}{x^2-2x-3}=\frac{2x-4}{(x-3)(x+1)}$$

Step 2: Set out the partial fractions using the factors of the denominator to make new denominators.

Let $\frac{2x-4}{(x-3)(x+1)}\equiv\frac{A}{x-3}+\frac{B}{x+1}$

Step 3: Add the fractions.

$$\frac{2x-4}{(x-3)(x+1)}\equiv\frac{A(x+1)+B(x-3)}{(x-3)(x+1)}$$

Step 4: Equate the numerators.

So

$$2x-4\equiv A(x+1)+B(x-3)$$

Step 5: To find A, substitute a value of x that will make the coefficient of B zero.

Substituting $x=3$,

$$2\times 3-4=A(3+1)+B(3-3)$$
$$2=4A$$
$$A=\tfrac{1}{2}$$

Step 6: To find B, substitute a value of x that will make the coefficient of A zero.

Substituting $x=-1$,

$$2\times -1-4=A(-1+1)+B(-1-3)$$
$$-6=-4B$$
$$B=\tfrac{3}{2}$$

Note: Use the equivalence sign $\equiv$ because the identity is true for all values of x. A and B are constants to be found.

Recall: Adding fractions (C3 Section 1.1).

Tip: Substituting $x=3$ makes the factor $(x-3)$ equal to zero.

Tip: A common error is to state $A=2$.

Tip: Substituting $x=-1$ makes the factor $(x+1)$ equal to zero.

Therefore

Step 7: Write out the partial fractions.

$$\frac{2x-4}{x^2-2x-3} \equiv \frac{1}{2(x-3)} + \frac{3}{2(x+1)}$$

Note:
$\frac{1}{2} \times \frac{1}{x-3} = \frac{1}{2(x-3)}$ and $\frac{3}{2} \times \frac{1}{x+1} = \frac{3}{2(x+1)}$.

In the above example the substitution method was used to find A and B. In the following example the method of equating coefficients is used. You will find it helpful to understand both methods when solving the more complicated examples later in the chapter.

Example 1.2 Express $\frac{5-6x}{(4x-3)(3x-2)}$ in partial fractions.

Step 1: Set out the partial fractions using the factors of the denominator to make new denominators.

Let $$\frac{5-6x}{(4x-3)(3x-2)} \equiv \frac{A}{4x-3} + \frac{B}{3x-2}$$

Step 2: Add the fractions.

$$\frac{5-6x}{(4x-3)(3x-2)} \equiv \frac{A(3x-2)+B(4x-3)}{(4x-3)(3x-2)}$$

Step 3: Equate the numerators.

So

$$5-6x \equiv A(3x-2)+B(4x-3)$$

Step 4: Expand the brackets.

$$5-6x \equiv 3Ax-2A+4Bx-3B$$

Tip:
Expanding the brackets will help you see the coefficients of the terms.

Step 5: Compare coefficients with the original numerator.

Equating coefficients of x,

$$-6 = 3A+4B \quad ①$$

Equating constant terms,

$$5 = -2A-3B \quad ②$$

Step 6: Solve the simultaneous equations for A and B.

① × 2 $\quad -12 = 6A+8B \quad$ ③

② × 3 $\quad 15 = -6A-9B \quad$ ④

③ + ④ $\quad 3 = -B$

$B = -3$

Tip:
Alternatively use the substitution method with $x = \frac{3}{4}$ and $x = \frac{2}{3}$. This will involve working with fractions accurately.

Substituting back into ①,

$$-6 = 3A-12$$
$$A = 2$$

Therefore

Step 7: Write out the partial fractions.

$$\frac{5-6x}{(4x-3)(3x-2)} \equiv \frac{2}{4x-3} - \frac{3}{3x-2}$$

Tip:
$+\frac{-3}{3x-2}$ is the same as $-\frac{3}{3x-2}$.

Example 1.3 **a** Express $\frac{3x^2+7x+8}{(x+1)(2x+1)(x-3)}$ in partial fractions.

b Hence differentiate $y = \frac{3x^2+7x+8}{(x+1)(2x+1)(x-3)}$ with respect to x.

Tip:
There are three factors in the original denominator, so there will be three partial fractions.

a

Step 1: Set out the partial fractions using the factors of the denominator to make new denominators.

Let $$\frac{3x^2+7x+8}{(x+1)(2x+1)(x-3)} \equiv \frac{A}{x+1} + \frac{B}{2x+1} + \frac{C}{x-3}$$

Step 2: Add the fractions.

$$\frac{3x^2+7x+8}{(x+1)(2x+1)(x-3)} \equiv \frac{A(2x+1)(x-3)+B(x+1)(x-3)+C(x+1)(2x+1)}{(x+1)(2x+1)(x-3)}$$

Step 3: Equate the numerators.

So

$$3x^2+7x+8 \equiv A(2x+1)(x-3)+B(x+1)(x-3)+C(x+1)(2x+1)$$

Step 4: To find A, substitute a value of x that will make the coefficients of B and C zero.

Substituting $x = -1$,

$$3(-1)^2 + 7(-1) + 8 = A(2 \times -1 + 1)(-1 - 3) + B \times 0 + C \times 0$$
$$4 = 4A$$
$$A = 1$$

Tip: Substitute $x = -1$ because then the factor $(x + 1)$ will equal zero.

Step 5: To find C, substitute a value of x that will make the coefficients of A and B zero.

Substituting $x = 3$,

$$3(3)^2 + 7 \times 3 + 8 = A \times 0 + B \times 0 + C(3 + 1)(2 \times 3 + 1)$$
$$56 = 28C$$
$$C = 2$$

Tip: Substitute $x = 3$ because then the factor $(x - 3)$ will equal zero.

Step 6: To find B, compare coefficients with the original numerator.

Equating coefficients of x^2,

$$3x^2 + \ldots \equiv A(2x^2 + \ldots) + B(x^2 + \ldots) + C(2x^2 + \ldots)$$

so $\quad 3 = 2A + B + 2C$

Substituting $A = 1$ and $C = 2$ gives

$$3 = 2 + B + 4$$
$$B = -3$$

Tip: You do not need to expand all the brackets, just the terms in x^2.

Tip: Alternatively you could substitute $x = -\frac{1}{2}$ but this leads to complex fraction work.

Therefore

Step 7: Write out the partial fractions.

$$\frac{3x^2 + 7x + 8}{(x + 1)(2x + 1)(x - 3)} = \frac{1}{x + 1} - \frac{3}{2x + 1} + \frac{2}{x - 3}$$

Tip: Check your answer by adding the partial fractions back together.

Step 1: Write your solution from **a** into a suitable format for differentiating.

b $$y = \frac{3x^2 + 7x + 8}{(x + 1)(2x + 1)(x - 3)} = \frac{1}{x + 1} - \frac{3}{2x + 1} + \frac{2}{x - 3}$$
$$= (x + 1)^{-1} - 3(2x + 1)^{-1} + 2(x - 3)^{-1}$$

Step 2: Differentiate with respect to x.

$$\frac{dy}{dx} = -(x + 1)^{-2} + 6(2x + 1)^{-2} - 2(x - 3)^{-2}$$
$$= \frac{6}{(2x + 1)^2} - \frac{1}{(x + 1)^2} - \frac{2}{(x - 3)^2}$$

Recall: $\frac{d}{dx} ax^n = nax^{n-1}$ (C1 Section 4.2). Use the chain rule to differentiate $-3(2x + 1)^{-1}$ (C3 Section 4.1).

Note: Although you could use the quotient rule on the original fraction, this format is easier for finding $\frac{d^2y}{dx^2}$ if this is required.

Fractions with a repeated linear factor in the denominator

A fraction such as $\frac{4x + 1}{x(2x - 1)^2}$ has three factors in the denominator so it will split into three partial fractions. These will be of the form $\frac{A}{x} + \frac{B}{2x - 1} + \frac{C}{(2x - 1)^2}$.

Note: If you have been taught the 'cover up' method, it does not work with fractions in this format.

Example 1.4 Express $\frac{4x + 1}{x(2x - 1)^2}$ in partial fractions.

Step 1: Set out the partial fractions using the factors of the denominator to make new denominators.

Let $$\frac{4x + 1}{x(2x - 1)^2} \equiv \frac{A}{x} + \frac{B}{2x - 1} + \frac{C}{(2x - 1)^2}$$

Step 2: Add the fractions.

$$\frac{4x + 1}{x(2x - 1)^2} \equiv \frac{A(2x - 1)^2 + Bx(2x - 1) + Cx}{x(2x - 1)^2}$$

Tip: Don't expand the brackets here; it's unnecessary and will make the substitutions harder.

Step 3: Equate the numerators.

So

$$4x + 1 \equiv A(2x - 1)^2 + Bx(2x - 1) + Cx$$

Step 4: To find A, substitute a value of x that will make the coefficients of B and C zero.

Substituting $x = 0$,

$$4 \times 0 + 1 = A(2 \times 0 - 1)^2 + B \times 0 + C \times 0$$
$$A = 1$$

Tip: One of the factors is x so make this zero.

Step 5: To find C, substitute a value of x that will make the coefficients of A and B zero.

Substituting $x = \frac{1}{2}$,

$$4 \times \tfrac{1}{2} + 1 = A \times 0 + B \times 0 + C \times \tfrac{1}{2}$$
$$3 = \tfrac{1}{2}C$$
$$C = 6$$

Tip: Substitute $x = \frac{1}{2}$ because then the factor $(2x - 1)$ equals zero.

Step 6: To find B, compare coefficients with the original numerator.

Equating coefficients of x^2,

$$0 = 4A + 2B$$
$$0 = 4 + 2B$$
$$B = -2$$

Tip: Alternatively you could substitute a third value of x or equate coefficients of x or constant terms.

Step 7: Write out the partial fractions.

Therefore $\dfrac{4x + 1}{x(2x - 1)^2} = \dfrac{1}{x} - \dfrac{2}{2x - 1} + \dfrac{6}{(2x - 1)^2}$

Improper fractions

An improper fraction is one in which the degree of the numerator is equal to or higher than the degree of the denominator.
To express an improper fraction in partial fractions first divide the numerator by the denominator to obtain a number and a proper fraction, using the skills you learnt in *Core 3*, then express the resulting proper fraction in partial fractions.

Recall: Long division of polynomials (C3 Section 1.1).

Example 1.5 Express $\dfrac{x^3 + 3x^2 + 1}{x^2 - x - 2}$ in partial fractions.

Step 1: Use long division of polynomials to express the fraction as a quotient and remainder.

$$\begin{array}{r} x + 4 \\ x^2 - x - 2 \overline{) x^3 + 3x^2 + 0x + 1} \\ \underline{x^3 - x^2 - 2x} \\ 4x^2 + 2x + 1 \\ \underline{4x^2 - 4x - 8} \\ 6x + 9 \end{array}$$

Tip: Don't forget to include $0x$ when setting up the division.

Tip: So $x^2 - x - 2$ divides in $x + 4$ times with a remainder of $6x + 9$.

So $\dfrac{x^3 + 3x^2 + 1}{x^2 - x - 2} = x + 4$ remainder $6x + 9$

i.e. $\dfrac{x^3 + 3x^2 + 1}{x^2 - x - 2} = x + 4 + \dfrac{6x + 9}{x^2 - x - 2}$

Note: After the long division the resulting proper fraction needs to be split into partial fractions.

Step 2: Factorise the denominator.

$$= x + 4 + \frac{6x + 9}{(x - 2)(x + 1)}$$

Step 3: Set out the partial fractions using the factors of the denominator to make new denominators.

Let $\dfrac{6x + 9}{(x - 2)(x + 1)} \equiv \dfrac{A}{x - 2} + \dfrac{B}{x + 1}$

Step 4: Add the fractions.

$$\frac{6x + 9}{(x - 2)(x + 1)} \equiv \frac{A(x + 1) + B(x - 2)}{(x - 2)(x + 1)}$$

Step 5: Equate the numerators.

So

$$6x + 9 \equiv A(x + 1) + B(x - 2)$$

Step 6: To find A, substitute a value of x that will make the coefficient of B zero.

Substituting $x = 2$,

$$6 \times 2 + 9 = A(2 + 1) + B(2 - 2)$$
$$21 = 3A$$
$$A = 7$$

Tip: Substitute $x = 2$ to make the factor $(x - 2)$ equal to zero.

Step 7: To find B, substitute a value of x that will make the coefficient of A zero.

Substituting $x = -1$,

$$6 \times -1 + 9 = A(-1 + 1) + B(-1 - 2)$$
$$3 = -3B$$
$$B = -1$$

Tip: Substitute $x = -1$ to make the factor $(x + 1)$ equal to zero.

Step 8: Write out the partial fractions.

Therefore $\dfrac{x^3 + 3x^2 + 1}{x^2 - x - 2} = x + 4 + \dfrac{7}{x - 2} - \dfrac{1}{x + 1}$

Tip: Don't forget to include the whole expression, not just the partial fractions.

SKILLS CHECK 1A: Partial fractions

1 Express the following in partial fractions:

a $\dfrac{4}{(x - 3)(x + 1)}$ **b** $\dfrac{x - 1}{(3x - 5)(x - 3)}$ **c** $\dfrac{4x - 13}{2x^2 + x - 6}$

2 The following fractions have repeated linear terms in their denominators. Express them in partial fractions.

a $\dfrac{1}{x^2(x - 1)}$ **b** $\dfrac{2x^2 + 3}{(x + 2)(x + 1)^2}$ **c** $\dfrac{4x^2 + 5x + 9}{(2x - 1)(x + 2)^2}$

3 Express the following improper fractions in partial fractions:

a $\dfrac{x^2 - 2x}{(x - 4)(x - 6)}$ **b** $\dfrac{x^3}{x^2 - 1}$ **c** $\dfrac{x^3 + 6x^2 - 1}{x^2 + 4x - 21}$

4 Express the following in partial fractions:

a $\dfrac{x + 27}{x^2 - 9}$ **b** $\dfrac{11 - 5x^2}{2 + x - x^2}$ **c** $\dfrac{3x}{(1 - x)(1 + x)^2}$

5 **a** Express $\dfrac{3}{(1 - 2x)(x + 1)}$ in partial fractions.

b Given the function $f(x) = \dfrac{3}{(1 - 2x)(x + 1)}$, $x \neq -1, x \neq \frac{1}{2}$, find the coordinates of the minimum point of the curve $y = f(x)$.

6 **a** Express $\dfrac{5x - 1}{1 - x^2}$ in partial fractions.

Given that $y = \dfrac{5x - 1}{1 - x^2}$,

b find $\dfrac{dy}{dx}$,

c find $\dfrac{d^2y}{dx^2}$ at the point where $x = 0$.

7 The function f is given by

$$f(x) = \frac{7 - 4x}{(2x + 3)(x - 5)}, \quad x \neq -\tfrac{3}{2}, x \neq 5$$

a Express $f(x)$ in partial fractions.

b Hence, or otherwise, prove that $f'(x) > 0$ for all values of x in the domain.

SKILLS CHECK 1A EXTRA is on the CD

Examination practice 1: Algebra and functions

1 Express $\dfrac{4x - 2}{x^3 - x}$ in partial fractions.

2 Express $\dfrac{x(2x - 5)}{(x + 2)(x - 1)^2}$ in partial fractions.

3 Show that $\dfrac{x^3 + 3x^2 + 10}{x^2 + 5x + 4}$ can be expressed in the form $A + Bx + \dfrac{C}{x + 1} + \dfrac{D}{x + 4}$, where A, B, C and D are constants to be found.

4 **a** Express $\dfrac{3x^2 - 9x + 7}{2x^2 - 7x + 6}$ in partial fractions.

b Hence find the x-coordinates of the stationary points on the curve $y = \dfrac{3x^2 - 9x + 7}{2x^2 - 7x + 6}$.

5 **a** Express $\dfrac{2}{x(x^2 - 1)}$ in partial fractions.

b Given the function $f(x) = \dfrac{2}{x(x^2 - 1)}$, $x \neq 0, x \neq \pm 1$, find $f'(2)$.

6 **a** Express $\dfrac{7}{(2x - 1)(x + 3)}$ in partial fractions.

Given that $y = \dfrac{7}{(2x - 1)(x + 3)}$,

b find $\dfrac{dy}{dx}$,

c find $\dfrac{d^2y}{dx^2}$.

7 The function f is given by

$$f(x) = \frac{3(x + 1)}{(x + 2)(x - 1)}, \quad x \in \mathbb{R}, x \neq -2, x \neq 1.$$

a Express $f(x)$ in partial fractions.

b Hence, or otherwise, prove that $f'(x) < 0$ for all values of x in the domain.

[Edexcel Jan 2003]

2 Coordinate geometry in the (x, y) plane

2.1 Parametric equations

Parametric equations of curves and conversion between Cartesian and parametric forms.

Up to now you have worked with Cartesian equations connecting two variables, usually x and y. Sometimes it is easier or more convenient to express x and y in terms of a third variable called a **parameter**.

For example,

$$x = t^2 + 1$$
$$y = t + 2$$

are the parametric equations of a curve.

Note: Here the parameter is t.

Example 2.1 **a** Draw the curve given by the parametric equations $x = t^2 + 1$, $y = t + 2$ for $-4 \leqslant t \leqslant 4$.

b Find a Cartesian equation of the curve.

Step 1: Draw up a table of values for x and y by substituting different values of t into x and y in the given interval.

a

t	−4	−3	−2	−1	0	1	2	3	4
$x = t^2 + 1$	17	10	5	2	1	2	5	10	17
$y = t + 2$	−2	−1	0	1	2	3	4	5	6

Tip: Use the patterns in the table to spot any errors.

Step 2: Plot the coordinates.

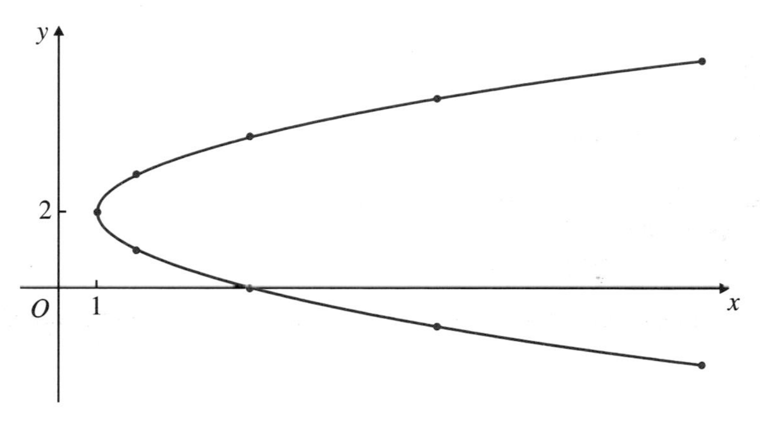

Tip: Each value of t gives a pair of coordinates, e.g. when $t = 2$, $x = 5$ and $y = 4$, so plot the point (5, 4).

Step 1: Rearrange one of the equations to find an expression for t.

b $x = t^2 + 1$, $y = t + 2$

$$y = t + 2$$

so

$$t = y - 2 \qquad ①$$

Since

$$x = t^2 + 1 \qquad ②$$

Step 2: Substitute the expression for t into the other parametric equation to eliminate t.

substituting ① into ② gives

$$x = (y - 2)^2 + 1$$

So a Cartesian equation of the curve is $x = (y - 2)^2 + 1$.

Tip: In this example it is easier to find an expression for t in terms of y than in terms of x.

Tip: Make sure your solution is in terms of x and y only.

Note: It is not always this straightforward to find the Cartesian form (see Example 2.3).

Example 2.2 A curve has parametric equations $x = 2t + 1$, $y = t^2 - 1$.

a Find the coordinates of the points of intersection of the curve with the axes.

b Find the points of intersection of the curve with the line $2y = x + 9$.

Step 1: Substitute $x = 0$ to find the value of t at the point where the curve cuts the y-axis.

a $x = 2t + 1$

When $x = 0$,

$$0 = 2t + 1$$
$$2t = -1$$
$$t = -\tfrac{1}{2}$$

Step 2: Substitute this value of t to find the y-coordinate.

When $t = -\frac{1}{2}$,

$$y = t^2 - 1 = (-\tfrac{1}{2})^2 - 1 = -\tfrac{3}{4}$$

So the curve cuts the y-axis at $(0, -\frac{3}{4})$.

Step 3: Substitute $y = 0$ to find the value of t at the point where the curve cuts the x-axis.

$y = t^2 - 1$

When $y = 0$,

$$0 = t^2 - 1$$
$$t^2 = 1$$
$$t = \pm 1$$

Tip: There will be two solutions when you take the square root.

Step 4: Substitute the values of t to find the x-coordinates.

When $t = 1$,

$$x = 2t + 1 = 2 \times 1 + 1 = 3$$

When $t = -1$,

$$x = 2t + 1 = 2 \times -1 + 1 = -1$$

So the curve cuts the x-axis at $(3, 0)$ and $(-1, 0)$.

Tip: Be careful to remember which coordinate you are looking for and substitute into the appropriate equation.

Step 1: Solve the equations of the line and the curve simultaneously by substituting both parametric equations into the equation of the line.

b $2y = x + 9$

At the point of intersection,

$$2(t^2 - 1) = (2t + 1) + 9$$
$$2t^2 - 2 = 2t + 10$$
$$2t^2 - 2t - 12 = 0$$
$$t^2 - t - 6 = 0$$
$$(t - 3)(t + 2) = 0$$
$$t = 3 \text{ or } -2$$

Tip: You are forming a quadratic equation in t. As usual you will need to collect terms and factorise to solve it.

Step 2: Substitute the values of t into the parametric equations to find the coordinates.

When $t = 3$,

$$x = 2t + 1 = 2 \times 3 + 1 = 7$$
$$y = t^2 - 1 = 3^2 - 1 = 8$$

When $t = -2$,

$$x = 2t + 1 = 2 \times -2 + 1 = -3$$
$$y = t^2 - 1 = (-2)^2 - 1 = 3$$

The points of intersection are $(7, 8)$ and $(-3, 3)$.

Curves with trigonometric parametric equations

To convert trigonometric parametric equations to Cartesian form, as before, you need to eliminate the parameter.

The identity $\cos^2\theta + \sin^2\theta \equiv 1$ is often useful for these problems.

Recall: Trigonometric identities (C2 Section 4.6).

Example 2.3 Find the Cartesian equation of each of the curves given by the following parametric equations:

a $x = 2\sin t - 3$, $y = \cos t + 1$ **b** $x = \cos\theta$, $y = \sin 2\theta$

Step 1: Rearrange the equations making $\sin t$ and $\cos t$ the subject.

a $x = 2\sin t - 3$

$$2\sin t = x + 3$$

$$\sin t = \frac{x+3}{2}$$

$y = \cos t + 1$

$$\cos t = y - 1$$

Since $\cos^2 t + \sin^2 t \equiv 1$,

Step 2: Eliminate t by squaring and adding.

$$(y-1)^2 + \left(\frac{x+3}{2}\right)^2 = 1$$

Tip:
$\cos t = y - 1$, so $\cos^2 t = (y-1)^2$
$\sin t = \frac{x+3}{2}$, so $\sin^2 t = \left(\frac{x+3}{2}\right)^2$

Note:
Try plotting this on a graphical calculator – it is the equation of an ellipse.

Step 1: Rewrite y in terms of $\sin\theta$ and $\cos\theta$.

b $y = \sin 2\theta$

$$= 2\sin\theta\cos\theta$$

Note:
Here the parameter is θ.

Step 2: Replace $\cos\theta$ by x.

Since $x = \cos\theta$,

$$y = 2\sin\theta \times x = 2x\sin\theta$$

Recall:
$\sin 2A = 2\sin A\cos A$
(C3 Section 2.3).

Step 3: Use trig identities to find an expression for $\sin\theta$ in terms of x.

Since $\cos^2\theta + \sin^2\theta \equiv 1$ and $x = \cos\theta$

$$x^2 + \sin^2\theta = 1$$

$$\sin^2\theta = 1 - x^2$$

$$\sin\theta = \sqrt{1 - x^2}$$

Tip:
You need to eliminate $\sin\theta$, so squaring is a good method to use.

Step 4: Rewrite y in terms of x alone.

Substituting into y gives

$$y = 2x\sin\theta$$

so $y = 2x\sqrt{1-x^2}$

Tip:
$\sqrt{1-x^2}$ cannot be simplified further.

Finding the area under a curve given its parametric equations

The area under a curve is given by $\int_{x=a}^{x=b} y\,dx$.

Recall:
Integration (C2 Section 7.2).

Using the chain rule, the area under a curve given in parametric form is

$$\text{Area} = \int_{t_1}^{t_2} y\frac{dx}{dt}\,dt$$

Note:
There are further examples of integration using parameters in Chapter 5.

Example 2.4 The diagram shows a sketch of part of the curve with parametric equations $x = (t+1)^2$, $y = t^3 + 1$, $t \geqslant 0$.

Note:
To find this area using the Cartesian equation of the curve you would need to work out $\int_1^9 ((\sqrt{x} - 1)^3 + 1)\,dx$.

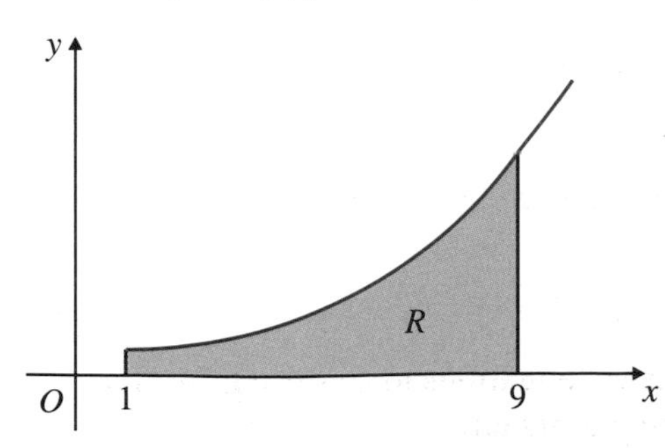

Find the area of the shaded region R.

Step 1: State the area formula in terms of x and y.

Area of $R = \int_1^9 y\,dx$

Step 2: Find the values of t that will be the new limits of integration.

When $x = 1$,

$$1 = (t+1)^2$$
$$t + 1 = \pm 1$$
$$t = 0$$

When $x = 9$,

$$9 = (t+1)^2$$
$$t + 1 = \pm 3$$
$$t = 2$$

Step 3: Find $\frac{dx}{dt}$.

$$x = (t+1)^2$$
$$\frac{dx}{dt} = \frac{d}{dt}(t+1)^2 = 2(t+1) = 2t + 2$$

Step 4: Use the chain rule and simplify.

$$\text{Area of } R = \int_0^2 y\frac{dx}{dt}\,dt = \int_0^2 (t^3 + 1) \times (2t + 2)\,dt$$
$$= \int_0^2 2t^4 + 2t^3 + 2t + 2\,dt$$

Step 5: Integrate term by term.

$$= \left[\frac{2t^5}{5} + \frac{t^4}{2} + t^2 + 2t\right]_0^2$$

Step 6: Substitute the limits and subtract.

$$= (\tfrac{64}{5} + 8 + 4 + 4) - (0)$$
$$= 28\tfrac{4}{5}$$

The area of R is $28\frac{4}{5}$ square units.

Tip: A common error is to forget to change the limits.

Tip: $t \neq -2$ as $t \geqslant 0$

Tip: $t \neq -4$ as $t \geqslant 0$

Recall: Differentiation using the chain rule (C3 Section 4.1).

Tip: Multiply out the brackets before attempting to integrate.

Recall: Definite integrals (C2 Section 7.1).

Tip: Show the substitution of both limits to get a mark for the method even if you make a later slip.

Example 2.5 The diagram shows a sketch of part of the curve with parametric equations $x = t^2 - 4$, $y = 3t + 2$.

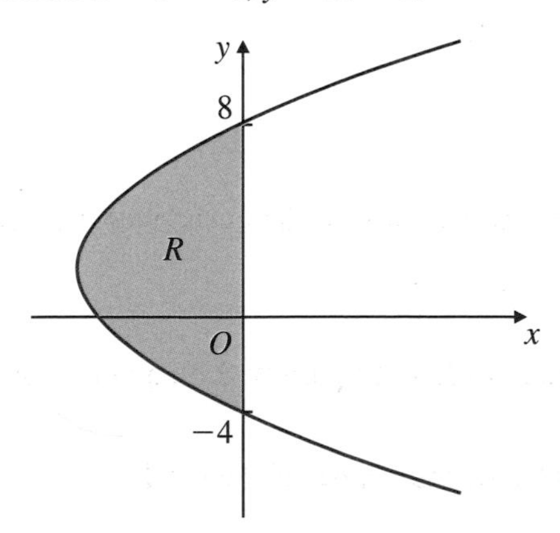

The shaded region, R, is enclosed between the curve and the y-axis. Find the area of R.

Note: This integral would be hard to find from the Cartesian equation $y = 3\sqrt{x+4} + 2$ since some of the area is above the x-axis and some below.

Step 1: Find the values of t that will be the limits of integration.

When $x = 0$,

$$0 = t^2 - 4$$
$$t^2 = 4$$
$$t = \pm 2$$

Tip: You could have found t when $y = -4$ and 8.

Tip: $t = \pm 2$ are your limits of integration.

***Step 2*: Find $\frac{dx}{dt}$.**

$$x = t^2 - 4$$

$$\frac{dx}{dt} = \frac{d}{dt}(t^2 - 4) = 2t$$

***Step 3*: Use the chain rule and simplify.**

$$\text{Area of } R = \int_{-2}^{2} y\frac{dx}{dt}\,dt = \int_{-2}^{2} (3t + 2) \times 2t\,dt$$

$$= \int_{-2}^{2} (6t^2 + 4t)\,dt$$

***Step 4*: Integrate term by term.**

$$= \Big[2t^3 + 2t^2\Big]_{-2}^{2}$$

***Step 5*: Substitute the limits and subtract.**

$$= (2 \times 2^3 + 2 \times 2^2) - (2 \times (-2)^3 + 2 \times (-2)^2)$$

$$= (16 + 8) - (-16 + 8)$$

$$= 24 - (-8)$$

$$= 32$$

The area of R is 32 square units.

Recall:
Definite integrals (C2 Section 7.1).

Tip:
Show the substitution of limits to get a mark for the method even if you make a later slip.

SKILLS CHECK **2A: Parametric equations**

1 Find the coordinates of any points where the following curves cut the y-axis:

a $x = t - 1, y = t^2 + 1$

b $x = 3\cos\theta, y = 4\sin\theta, 0 \leqslant \theta < 2\pi$

2 Find the coordinates of any points where the following curves cut the x-axis:

a $x = 2t, y = 3 - t$

b $x = 3t + 2, y = t^2 - 1$

3 A curve has parametric equations $x = \frac{at^2}{2}$, $y = a(t - 1)$, where a is a positive constant.

Given that the curve passes through the point (2, 0), find the value of a.

4 Find a Cartesian equation for each of the following curves:

a $x = t + 3, y = \frac{1}{t}, t \neq 0$

b $x = \cos\theta, y = \tan\theta$

c $x = \sec\theta, y = 3\tan\theta$

5 Find the coordinates of the points of intersection of the curve with parametric equations $x = \frac{3}{1 + t^2}$, $y = \frac{t}{1 + t^2}$ and the line $x + y = 1$.

6 **a** Find the Cartesian equation of the circle with parametric equations $x = 3\cos\theta + 2$, $y = 3\sin\theta - 4$.

b Find the radius and the coordinates of the centre of the circle.

7 The diagram shows a sketch of part of the curve with parametric equations $x = t - 1, y = 2t^2 + 1$.
Find the area of the shaded region R.

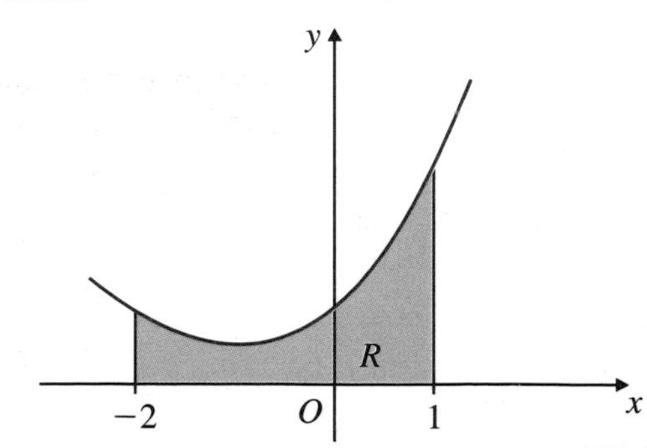

8 The diagram shows a sketch of part of the curve with parametric equations $x = t^2$, $y = t(4 - t)$. Find the area of the shaded region R.

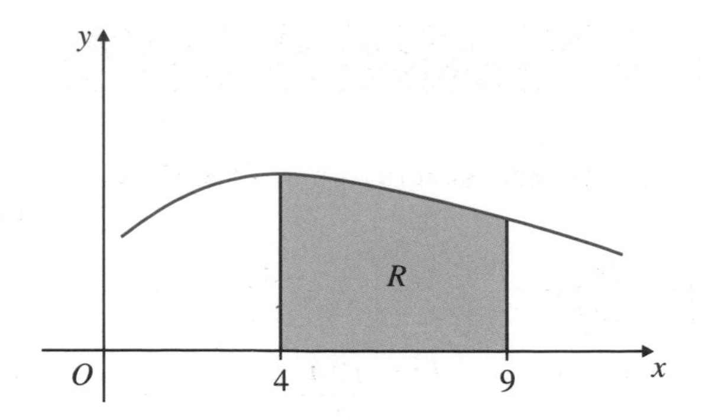

SKILLS CHECK **2A EXTRA is on the CD**

Examination practice 2: Coordinate geometry in the (x, y) plane

1 A curve has parametric equations $x = ct$, $y = \frac{c}{t}$, $t > 0$, where c is a positive constant.

Given that the curve passes through the point (9, 1), find the value of c.

2 A curve has parametric equations $x = \sin t$, $y = \cos 2t$, $0 \leq t < 2\pi$.

a Find the exact coordinates of the points of intersection of the curve with the axes.

b Find a Cartesian equation of the curve.

3 **a** Find a Cartesian equation for the curve with parametric equations

$$x = t - 1, \qquad y = t^3 + t^2$$

b Find the coordinates of the point at which the curve crosses the x-axis.

4 A curve has parametric equations $x = 2kt^2$, $y = k(8t - t^4)$, where k is a positive constant. Given that the curve passes through the point (24, 0), find the value of k.

5 Find the exact coordinates of the points of intersection of the curve with parametric equations $x = \sin\theta - 1$, $y = \sqrt{3}\cos\theta + 1$, $0 \leq \theta < 2\pi$, and the line $y = x + 2$.

6 **a** Find the Cartesian equation of the circle with parametric equations $x = 4\cos t - 1$, $y = 4\sin t + 5$.

b Find the radius and the coordinates of the centre of the circle.

7 A curve has parametric equations $x = 2t^2$, $y = 4t - 1$. Find an equation of the line joining the points on the curve where $t = -2$ and $t = 1$.

8 The diagram shows part of the curve with parametric equations $x = t^2$, $y = 4t - t^3$. The curve forms a loop between the points (0, 0) and (4, 0).

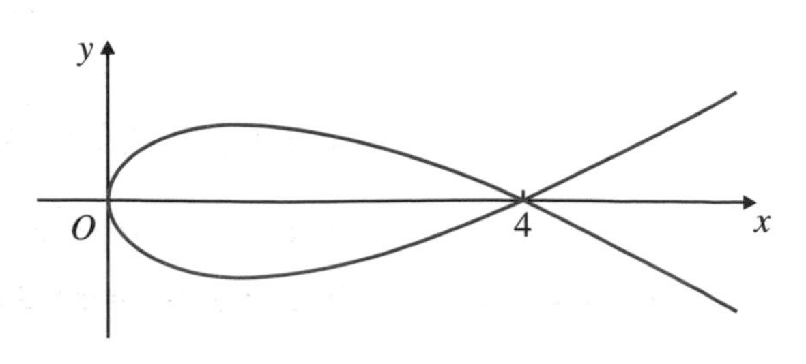

a The loop is traced out by values of t between t_1 and t_2. Find the values of t_1 and t_2.

b Find the area of the region enclosed by the loop of the curve.

3 Sequences and series

3.1 Binomial series

Binomial series for any rational n.

In *Core 2* you used the binomial expansion to expand $(1+x)^n$ for any positive integer n:

$$(1+x)^n = 1 + \binom{n}{1}x + \binom{n}{2}x^2 + \binom{n}{3}x^3 + \cdots + \binom{n}{r}x^r + \cdots$$

$$= 1 + nx + \frac{n(n-1)}{2!}x^2 + \frac{n(n-1)(n-2)}{3!}x^3 + \cdots$$

$$+ \frac{n(n-1)\ldots(n-r+1)}{r!}x^r + \cdots$$

When n is a positive integer the series is finite and gives the exact value of $(1+x)^n$.

The expansion can also be used for fractional and negative values of n. However, the series will be infinite since none of the factors of the coefficients will ever be zero.

When $|x| < 1$ the series is a valid approximation to $(1+x)^n$ as it will be convergent.

Recall: Binomial expansion (C2 Section 3.2).

Tip: $\binom{n}{2}$ means $\frac{n(n-1)}{2!}$. Don't confuse it with $\frac{n}{2}$.

Recall: $r! = r(r-1)(r-2)\ldots \times 3 \times 2 \times 1$

Note: $|x| < 1$ can be written as $-1 < x < 1$.

Example 3.1 Find the binomial expansion of the following in ascending powers of x up to and including the term in x^3. State the range of values of x for which each expansion is valid.

a $(1+x)^{-3}$ **b** $\sqrt{1-2x}$ **c** $\frac{1}{(1+3x)^{\frac{2}{3}}}$

Step 1: Substitute into the binomial expansion.
Step 2: Simplify the terms.
Step 3: State when the expansion is valid.

a $(1+x)^{-3} = 1 + (-3)x + \frac{(-3)(-4)}{2!}x^2 + \frac{(-3)(-4)(-5)}{3!}x^3 + \cdots$

$= 1 - 3x + 6x^2 - 10x^3 + \cdots$

The expansion is valid when $|x| < 1$.

Tip: Replace n by -3 in the expansion.

Note: The coefficients are $1, -3, 6, -10\ldots$

Step 1: Write in index form.
Step 2: Substitute into the binomial expansion.
Step 3: Simplify the terms.
Step 4: State when the expansion is valid.

b $\sqrt{1-2x} = (1-2x)^{\frac{1}{2}}$

$$= 1 + (\tfrac{1}{2})(-2x) + \frac{(\frac{1}{2})(-\frac{1}{2})}{2!}(-2x)^2 + \frac{(\frac{1}{2})(-\frac{1}{2})(-\frac{3}{2})}{3!}(-2x)^3 + \cdots$$

$$= 1 + (\tfrac{1}{2})(-2x) + \frac{(\frac{1}{2})(-\frac{1}{2})}{2!}(-2x)(-2x)$$

$$+ \frac{(\frac{1}{2})(-\frac{1}{2})(-\frac{3}{2})}{3!}(-2x)(-2x)(-2x) + \cdots$$

$$= 1 + (\tfrac{1}{2})(-2x) + \frac{(\frac{1}{2})(-\frac{1}{2})}{2!}(4x^2) + \frac{(\frac{1}{2})(-\frac{1}{2})(-\frac{3}{2})}{3!}(-8x^3) + \cdots$$

$$= 1 - x - \tfrac{1}{2}x^2 - \tfrac{1}{2}x^3 + \cdots$$

The expansion is valid when $|-2x| < 1$, i.e. when $|x| < \frac{1}{2}$.

Tip: Replace n by $\frac{1}{2}$ and x by $-2x$ in the expansion.

Tip: Don't forget to square the -2 as well as the x: $(-2x)^2 = 4x^2$ *not* $-2x^2$. It can help to write this out in full.

Note: The coefficients are $1, -1, -\frac{1}{2}, -\frac{1}{2}, \ldots$

Tip: $-1 < -2x < 1$, so $\frac{1}{2} > x > -\frac{1}{2}$ i.e. $-\frac{1}{2} < x < \frac{1}{2}$ so $|x| < \frac{1}{2}$

Step 1: Write in index form.

c $\dfrac{1}{(1+3x)^{\frac{2}{3}}} = (1+3x)^{-\frac{2}{3}}$

Step 2: Substitute into the binomial expansion.

$= 1 + (-\tfrac{2}{3})(3x) + \dfrac{(-\frac{2}{3})(-\frac{5}{3})}{2!}(3x)^2 + \dfrac{(-\frac{2}{3})(-\frac{5}{3})(-\frac{8}{3})}{3!}(3x)^3 + \cdots$

Tip: Replace n by $-\frac{2}{3}$ and x by $3x$ in the expansion.

Step 3: Simplify the terms.

$= 1 + (-\tfrac{2}{3})(3x) + \dfrac{(-\frac{2}{3})(-\frac{5}{3})}{2!}(3x)(3x)$

$+ \dfrac{(-\frac{2}{3})(-\frac{5}{3})(-\frac{8}{3})}{3!}(3x)(3x)(3x) + \cdots$

$= 1 + (-\tfrac{2}{3})(3x) + \dfrac{(-\frac{2}{3})(-\frac{5}{3})}{2!}(9x^2) + \dfrac{(-\frac{2}{3})(-\frac{5}{3})(-\frac{8}{3})}{3!}(27x^3) + \cdots$

$= 1 - 2x + 5x^2 - \frac{40}{3}x^3 + \cdots$

Note: The coefficients are $1, -2, 5, -\frac{40}{3}, \ldots$

Step 4: State when the expansion is valid.

The expansion is valid when $|3x| < 1$, i.e. when $|x| < \frac{1}{3}$.

Tip: $-1 < 3x < 1$, so $-\frac{1}{3} < x < \frac{1}{3}$

Approximations

The binomial expansion is often used to make approximations.

Example 3.2 **a** Find the binomial expansion of $\sqrt{1-4x}$ in ascending powers of x up to and including the term in x^3.

b By substituting $x = 0.02$ in your expansion, find an approximation to $\sqrt{23}$, giving your answer to five significant figures.

Step 1: Write in index form.

a $\sqrt{1-4x} = (1-4x)^{\frac{1}{2}}$

Step 2: Substitute into the binomial expansion.

$= 1 + (\tfrac{1}{2})(-4x) + \dfrac{(\frac{1}{2})(-\frac{1}{2})}{2!}(-4x)^2 + \dfrac{(\frac{1}{2})(-\frac{1}{2})(-\frac{3}{2})}{3!}(-4x)^3 + \cdots$

Tip: Replace n by $\frac{1}{2}$ and x by $-4x$ in the expansion.

Step 3: Simplify the terms.

$= 1 + (\tfrac{1}{2})(-4x) + \dfrac{(\frac{1}{2})(-\frac{1}{2})}{2!}(16x^2) + \dfrac{(\frac{1}{2})(-\frac{1}{2})(-\frac{3}{2})}{3!}(-64x^3) + \cdots$

Tip: If you need to you can continue to include the extra line of working showing $(-4x)^2 = (-4x)(-4x)$ and $(-4x)^3 = (-4x)(-4x)(-4x)$.

Step 4: State when the expansion is valid.

$= 1 - 2x - 2x^2 - 4x^3 + \cdots$

The expansion is valid when $|-4x| < 1$, i.e. when $|x| < \frac{1}{4}$.

Step 1: Substitute the given value into $\sqrt{1-4x}$.

b When $x = 0.02$,

$\sqrt{1 - 4 \times 0.02} = \sqrt{0.92}$

Step 2: Simplify the surds.

$= \sqrt{\dfrac{92}{100}}$

$= \sqrt{\dfrac{4 \times 23}{100}}$

$= \dfrac{2\sqrt{23}}{10}$

Tip: Rewrite 0.92 as $\frac{92}{100}$.

Recall: Manipulation of surds (C1 Section 1.2).

Step 3: Substitute the given value into your expansion.

$\sqrt{1 - 4 \times 0.02} \approx 1 - 2 \times 0.02 - 2(0.02)^2 - 4(0.02)^3$

$\approx 1 - 0.04 - 0.0008 - 0.000\,032$

$\approx 0.959\,168$

Note: $x = 0.02$ is valid since $0.02 < \frac{1}{4}$.

Step 4: Equate the values and rearrange to find the required value.

So $\dfrac{2\sqrt{23}}{10} \approx 0.959\,168$

$\sqrt{23} \approx \dfrac{0.959\,168 \times 10}{2}$

$\sqrt{23} \approx 4.795\,84 \approx 4.7958$ (5 s.f.)

Tip: You must show your method but you can check your answer by finding $\sqrt{23}$ on your calculator.

Expanding $(a + bx)^n$

The binomial expansion of $(1 + x)^n$ can be used to expand expressions of the form $(a + bx)^n$. You must take out a factor of a^n to turn the expression into the required format.

Example 3.3 For each of the following expressions find the binomial expansion in ascending powers of x up to and including the term in x^3. State the range of values of x for which each expansion is valid.

a $(8 + 3x)^{\frac{1}{3}}$ **b** $\dfrac{1 - x}{(2 + x)^2}$

> **Tip:** A common mistake is to forget that 8 should be to the power $\frac{1}{3}$ too.

Step 1: Take out a factor.

a $(8 + 3x)^{\frac{1}{3}} = [8(1 + \frac{3}{8}x)]^{\frac{1}{3}} = 8^{\frac{1}{3}}(1 + \frac{3}{8}x)^{\frac{1}{3}}$

Step 2: Substitute into the binomial expansion.

$$= 8^{\frac{1}{3}}\left\{1 + (\tfrac{1}{3})(\tfrac{3}{8}x) + \frac{(\frac{1}{3})(-\frac{2}{3})}{2!}(\tfrac{3}{8}x)^2 + \frac{(\frac{1}{3})(-\frac{2}{3})(-\frac{5}{3})}{3!}(\tfrac{3}{8}x)^3 + \cdots\right\}$$

> **Tip:** Replace n by $\frac{1}{3}$ and x by $\frac{3}{8}x$.

Step 3: Simplify the terms.

$$= 2\left\{1 + (\tfrac{1}{3})(\tfrac{3}{8}x) + \frac{(\frac{1}{3})(-\frac{2}{3})}{2!}(\tfrac{9}{64}x^2) + \frac{(\frac{1}{3})(-\frac{2}{3})(-\frac{5}{3})}{3!}(\tfrac{27}{512}x^3) + \cdots\right\}$$

> **Note:** $8^{\frac{1}{3}} = \sqrt[3]{8} = 2$

Step 4: Multiply through by the factor.

$$= 2\{1 + \tfrac{1}{8}x - \tfrac{1}{64}x^2 + \tfrac{5}{1536}x^3 + \cdots\}$$

$$= 2 + \tfrac{1}{4}x - \tfrac{1}{32}x^2 + \tfrac{5}{768}x^3 + \cdots$$

> **Tip:** Don't forget to multiply all the terms by 2.

Step 5: State when the expansion is valid.

The expansion is valid when $|\frac{3}{8}x| < 1$, i.e. when $|x| < \frac{8}{3}$.

> **Tip:** Make sure you use $\frac{3}{8}x$ and not $3x$ here.

Step 1: Write the denominator in index form.

b $\dfrac{1 - x}{(2 + x)^2} = (1 - x)(2 + x)^{-2}$

Step 2: Take out a factor.

$= (1 - x)[2(1 + \frac{1}{2}x)]^{-2}$

$= (1 - x)(2^{-2})(1 + \frac{1}{2}x)^{-2}$

> **Tip:** A common mistake is to forget that 2 should be to the power -2 as well.

> **Note:** $2^{-2} = \dfrac{1}{2^2} = \dfrac{1}{4}$

Step 3: Substitute into the binomial expansion.

$$= \tfrac{1}{4}(1 - x)\left\{1 + (-2)(\tfrac{1}{2}x) + \frac{(-2)(-3)}{2!}(\tfrac{1}{2}x)^2 + \frac{(-2)(-3)(-4)}{3!}(\tfrac{1}{2}x)^3 + \cdots\right\}$$

> **Tip:** Replace n by -2 and x by $\frac{1}{2}x$.

Step 4: Simplify the terms.

$$= \tfrac{1}{4}(1 - x)\left\{1 + (-2)(\tfrac{1}{2}x) + \frac{(-2)(-3)}{2!}(\tfrac{1}{4}x^2) + \frac{(-2)(-3)(-4)}{3!}(\tfrac{1}{8}x^3) + \cdots\right\}$$

Step 5: Expand the brackets.

$$= \tfrac{1}{4}(1 - x)\{1 - x + \tfrac{3}{4}x^2 - \tfrac{1}{2}x^3 + \cdots\}$$

Step 6: Collect terms.

$$= \tfrac{1}{4}\{1 - x + \tfrac{3}{4}x^2 - \tfrac{1}{2}x^3 - x + x^2 - \tfrac{3}{4}x^3 + \cdots\}$$

$$= \tfrac{1}{4}\{1 - 2x + \tfrac{7}{4}x^2 - \tfrac{5}{4}x^3 + \cdots\}$$

> **Tip:** You are only asked for four terms so you can ignore the last one when you expand the brackets.

Step 7: Multiply through by the factor.

$$= \tfrac{1}{4} - \tfrac{1}{2}x + \tfrac{7}{16}x^2 - \tfrac{5}{16}x^3 + \cdots$$

> **Tip:** It's easier to expand the brackets and then multiply through by $\frac{1}{4}$.

Step 8: State when the expansion is valid.

The expansion is valid when $|\frac{1}{2}x| < 1$, i.e. when $|x| < 2$.

Using partial fractions

More complex expressions can often be simplified using partial fractions and then expanded one part at a time.

Example 3.4 **a** Express $\dfrac{1}{(1-x)(1+2x)}$ in partial fractions.

b Hence find the first four terms of the binomial expansion of $\dfrac{1}{(1-x)(1+2x)}$ in ascending powers of x.

c State the range of values of x for which the expansion is valid.

Note: Part **b** could be rewritten as $(1-x)^{-1}(1+2x)^{-1}$ and expanded, but the use of the word 'hence' in part **b** means that you are expected to use your result from part **a**.

Recall: Partial fractions (Section 1.1).

Step 1: Set out the partial fractions using the factors of the denominator to make new denominators.

a Let $$\frac{1}{(1-x)(1+2x)} \equiv \frac{A}{1-x} + \frac{B}{1+2x}$$

Step 2: Add the fractions.

$$\frac{1}{(1-x)(1+2x)} \equiv \frac{A(1+2x)+B(1-x)}{(1-x)(1+2x)}$$

Step 3: Equate the numerators.

So $$1 \equiv A(1+2x) + B(1-x)$$

Step 4: To find A, substitute a value of x that will make the coefficient of B zero.

Substituting $x = 1$,
$$1 = A(1 + 2\times 1) + B \times 0$$
$$1 = 3A$$
$$A = \tfrac{1}{3}$$

Step 5: To find B, substitute a value of x that will make the coefficient of A zero.

Substituting $x = -\frac{1}{2}$,
$$1 = A\times 0 + B(1-(-\tfrac{1}{2}))$$
$$1 = \tfrac{3}{2}B$$
$$B = \tfrac{2}{3}$$

Step 6: Write out the solution.

Therefore $$\frac{1}{(1-x)(1+2x)} \equiv \frac{1}{3(1-x)} + \frac{2}{3(1+2x)}$$

Step 1: Write the denominator in index form.

b $$\frac{1}{(1-x)(1+2x)} \equiv \frac{1}{3(1-x)} + \frac{2}{3(1+2x)}$$
$$\equiv \tfrac{1}{3}(1-x)^{-1} + \tfrac{2}{3}(1+2x)^{-1}$$

Tip: Use your answer from part **a**.

Step 2: Expand the first fraction by substituting into the binomial expansion.

$$\tfrac{1}{3}(1-x)^{-1} = \tfrac{1}{3}\left\{1 + (-1)(-x) + \frac{(-1)(-2)}{2!}(-x)^2 + \frac{(-1)(-2)(-3)}{3!}(-x)^3 + \cdots\right\}$$

Tip: Replace n by -1 and x by $-x$.

Step 3: Simplify the terms.

$$= \tfrac{1}{3}\{1 + x + x^2 + x^3 + \cdots\}$$
$$= \tfrac{1}{3} + \tfrac{1}{3}x + \tfrac{1}{3}x^2 + \tfrac{1}{3}x^3 + \cdots$$

Step 4: Expand the second fraction by substituting into the binomial expansion.

$$\tfrac{2}{3}(1+2x)^{-1} = \tfrac{2}{3}\left\{1 + (-1)(2x) + \frac{(-1)(-2)}{2!}(2x)^2 + \frac{(-1)(-2)(-3)}{3!}(2x)^3 + \cdots\right\}$$

Tip: Replace n by -1 and x by $2x$.

Step 5: Simplify the terms.

$$= \tfrac{2}{3}\{1 - 2x + 4x^2 - 8x^3 + \cdots\}$$
$$= \tfrac{2}{3} - \tfrac{4}{3}x + \tfrac{8}{3}x^2 - \tfrac{16}{3}x^3 + \cdots$$

Step 6: Combine the expansions.

$$\frac{1}{(1-x)(1+2x)} = \tfrac{1}{3}(1-x)^{-1} + \tfrac{2}{3}(1+2x)^{-1}$$
$$= \tfrac{1}{3} + \tfrac{1}{3}x + \tfrac{1}{3}x^2 + \tfrac{1}{3}x^3 + \tfrac{2}{3} - \tfrac{4}{3}x + \tfrac{8}{3}x^2 - \tfrac{16}{3}x^3 + \cdots$$
$$= 1 - x + 3x^2 - 5x^3 + \cdots$$

Tip: You could try checking your answer by expanding $(1-x)^{-1}(1+2x)^{-1}$.

Step 1: State when each expansion is valid.

c The expansion of $\frac{1}{3}(1-x)^{-1}$ is valid when $|-x| < 1$, i.e. $|x| < 1$.

The expansion of $\frac{2}{3}(1+2x)^{-1}$ is valid when $|2x| < 1$, i.e. $|x| < \frac{1}{2}$.

Tip:
Use an open circle to show that x cannot equal the endpoints of the set of values.

Step 2: Decide which values of x satisfy both inequalities.

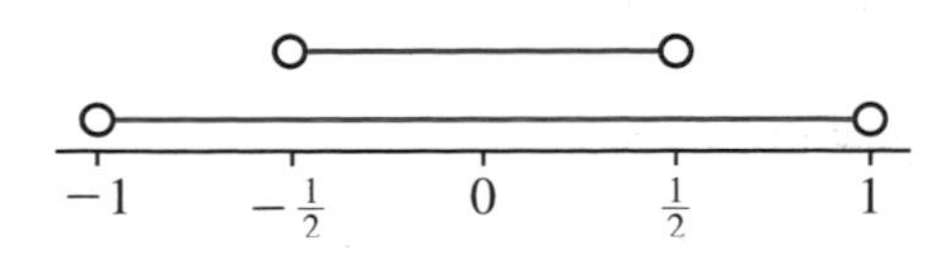

Both expansions are valid when $|x| < \frac{1}{2}$.

SKILLS CHECK **3A: Binomial series**

1 For each of the following expressions find the binomial expansion in ascending powers of x up to and including the term in x^3. State the range of values of x for which each expansion is valid.

a $\dfrac{1}{(1-x)^2}$ **b** $\sqrt[3]{1-9x}$ **c** $\dfrac{1}{(1-4x)^{\frac{3}{2}}}$

2 For each of the following expressions find the binomial expansion in ascending powers of x up to and including the term in x^2. State the ranges of values of x for which each expansion is valid.

a $\dfrac{1}{(2x-1)^3}$ **b** $(4+x)^{-\frac{3}{2}}$

3 Find the first four terms in the expansion of $\dfrac{2+x}{(1+x)^2}$ in ascending powers of x.

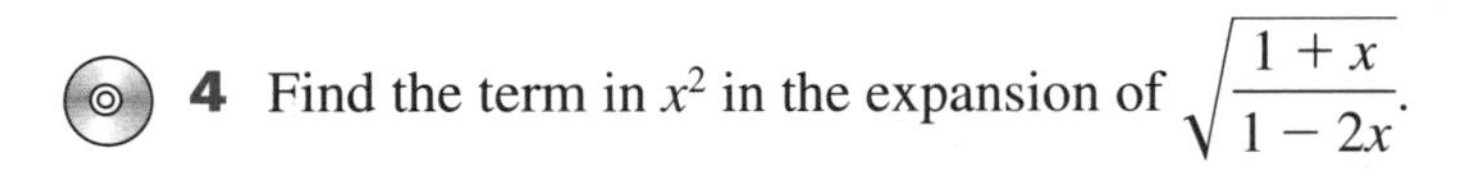

4 Find the term in x^2 in the expansion of $\sqrt{\dfrac{1+x}{1-2x}}$.

5 Find the binomial expansion of $\sqrt{1+x}$ in ascending powers of x up to and including the term in x^3. By substituting $x = 0.08$ into your expansion, find an approximation to $\sqrt{3}$, giving your answer to five significant figures.

6 **a** Express $\dfrac{2x^2-5x}{(2+x)(1-x)^2}$ in partial fractions.

b Hence find the first three terms of the binomial expansion of $\dfrac{2x^2-5x}{(2+x)(1-x)^2}$ in ascending powers of x.

c State the range of values of x for which the expansion is valid.

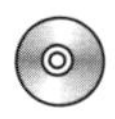

7 In the expansion of $(1+ax)^n$, in ascending powers of x, the coefficient of x is 2 and the coefficient of x^3 is twice the coefficient of x^2.

a Find a and n.

b Find the coefficient of x^3.

c State the values of x for which the expansion is valid.

SKILLS CHECK **3A EXTRA is on the CD**

Examination practice 3: Sequences and series

1 **a** Use the binomial series to expand $(3 + 2x)^{-3}$, in ascending powers of x, up to and including the term in x^3. Give each coefficient as a simplified fraction.

b State the values of x for which the expansion is valid.

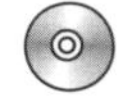

2 Given that

$$\frac{(2x - 3)}{(1 - x)(2 - x)} \equiv \frac{A}{1 - x} + \frac{B}{2 - x}$$

a find the values of the constants A and B.

b Hence, or otherwise, find the series expansion in ascending powers of x, up to and including the term in x^3, of $\frac{(2x - 3)}{(1 - x)(2 - x)}$.

c State the values of x for which the expansion is valid.

3 When $(1 + kx)^n$ is expanded as a series in ascending powers of x, the first three terms are $1 - x + \frac{3}{2}x^2$.

a Find the value of k and the value of n.

b Find the coefficient of x^3.

c State the set of values of x for which the expansion is valid.

4 **a** Expand $(1 + 3x)^{-2}$, $|x| < \frac{1}{3}$, in ascending powers of x up to and including the term in x^3, simplifying each term.

b Hence, or otherwise, find the first three terms in the expansion of $\frac{x + 4}{(1 + 3x)^2}$ as a series in ascending powers of x. [Edexcel Jan 2003]

5 Given that

$$\frac{3 - 4x}{(1 - x)(1 - 2x)} \equiv \frac{A}{1 - x} + \frac{B}{1 - 2x}$$

a find the values of the constants A and B.

b Hence, or otherwise, find the series expansion in ascending powers of x, up to and including the term in x^2, of

$$\frac{3 - 4x}{(1 - x)(1 - 2x)}$$

c State the set of values of x for which the expansion is valid.

6 The binomial expansion of $(8 + x)^{\frac{1}{3}}$ in ascending powers of x, as far as the term in x^2, is

$$(8 + x)^{\frac{1}{3}} = 2 + px + qx^2 + \ldots, \quad |x| < 8.$$

a Determine the values of the constants p and q.

b Use the expression $2 + px + qx^2$, and your values of p and q, to obtain an estimate for $\sqrt[3]{15}$, giving your answer to 3 significant figures.

c Find the percentage error involved in using this estimate. [London May 1995]

7 In the binomial expansion, in ascending powers of x, of $(1 + ax)^n$, where a and n are constants, the coefficient of x is 15. The coefficient of x^2 and of x^3 are equal.

a Find the value of a and the value of n.

b Find the coefficient of x^3. [Edexcel May 2005]

8 $$f(x) = \frac{1}{\sqrt{(1 - x)}} - \sqrt{(1 + x)}, \quad -1 < x < 1.$$

a Find the series expansion of $f(x)$, in ascending powers of x, up to and including the term in x^3.

b Hence, or otherwise, prove that the function f has a minimum at the origin. [Edexcel Jan 2005]

9 Given that

$$\frac{10(2 - 3x)}{(1 - 2x)(2 + x)} \equiv \frac{A}{1 - 2x} + \frac{B}{2 + x},$$

a find the values of the constants A and B.

b Hence, or otherwise, find the series expansion in ascending powers of x, up to and including the term in x^3, of $\frac{10(2 - 3x)}{(1 - 2x)(2 + x)}$, for $|x| < \frac{1}{2}$. [Edexcel Jan 2002]

4 Differentiation

4.1 Differentiation of implicit and parametric functions

Differentiation of simple functions defined implicitly or parametrically.

Implicit functions

In *Core 3* the functions that you differentiated were of the form $y = f(x)$, where y was given explicitly in terms of x, or $x = f(y)$, where x was given explicitly in terms of y.

Sometimes a function in two variables is defined **implicitly**, and neither variable is given explicitly in terms of the other.

You will need to be able to differentiate implicit functions such as $3x^2 + y^2 = 9$ and $x^2y + y^3 = 2x$.

To differentiate an implicit function:

- differentiate term by term with respect to x
- use the chain rule to differentiate any terms in y

Recall: The chain rule (C3 Section 4.1).

For example, to differentiate y^2 with respect to x use the chain rule:

$$\frac{d}{dx}(y^2) = \frac{d}{dy}(y^2) \times \frac{dy}{dx}$$
$$= 2y\frac{dy}{dx}$$

Tip: When you have understood implicit differentiation you will be able to show far less working.

Example 4.1 Find $\frac{dy}{dx}$ in terms of x and y for each of the following:

a $3x^2 + y^2 = 9$

b $x^2y + y^3 = 2x$

Step 1: Differentiate term by term with respect to x.

a $3x^2 + y^2 = 9$

Differentiating with respect to x,

Tip: Don't forget any terms on the right-hand side.

$$\frac{d}{dx}(3x^2) + \frac{d}{dx}(y^2) = \frac{d}{dx}(9)$$

Step 2: Use the chain rule to differentiate y^2 with respect to x.

$$6x + \frac{d}{dy}(y^2)\frac{dy}{dx} = 0$$

$$6x + 2y\frac{dy}{dx} = 0$$

Step 3: Rearrange to make $\frac{dy}{dx}$ the subject.

Tip: Divide through by the coefficient of $\frac{dy}{dx}$.

$$2y\frac{dy}{dx} = -6x$$

$$\frac{dy}{dx} = -\frac{6x}{2y} = -\frac{3x}{y}$$

Step 1: Differentiate term by term with respect to x.

b $x^2y + y^3 = 2x$

Differentiating with respect to x,

$$\frac{d}{dx}(x^2y) + \frac{d}{dx}(y^3) = \frac{d}{dx}(2x)$$

Step 2: Use the product rule to differentiate x^2y with respect to x.

$$\left(\frac{d}{dx}(x^2) \times y + x^2 \times \frac{d}{dx}(y)\right) + \frac{d}{dx}(y^3) = 2$$

Step 3: Use the chain rule to differentiate y and y^3 with respect to x.

$$2xy + x^2\frac{d}{dy}(y)\frac{dy}{dx} + \frac{d}{dy}(y^3)\frac{dy}{dx} = 2$$

$$2xy + x^2\frac{dy}{dx} + 3y^2\frac{dy}{dx} = 2$$

$$(x^2 + 3y^2)\frac{dy}{dx} = 2 - 2xy$$

Step 4: Rearrange to make $\frac{dy}{dx}$ the subject.

$$\frac{dy}{dx} = \frac{2 - 2xy}{x^2 + 3y^2}$$

Recall:
Product rule (C3 Section 4.2).

Tip:
$\frac{d}{dy}(y) = 1$

Tip:
Collect all terms containing $\frac{dy}{dx}$ on one side and factorise.

Example 4.2 A circle has centre (2, 3) and radius $\sqrt{5}$.

a State the equation of the circle.

b Use implicit differentiation to find an equation of the normal to the circle at (1, 5).

Note:
Although you learnt another method for answering **b** in C2, this question tells you to use implicit differentiation.

Step 1: Use the general equation of a circle.

a The circle with centre (2, 3) and radius $\sqrt{5}$ has equation
$(x - 2)^2 + (y - 3)^2 = 5$

Recall:
A circle, centre (a, b), radius r, has equation
$(x - a)^2 + (y - b)^2 = r^2$
(C2 Section 2.1).

Step 2: Differentiate term by term with respect to x.

b $(x - 2)^2 + (y - 3)^2 = 5$

Differentiating with respect to x,

$$\frac{d}{dx}(x - 2)^2 + \frac{d}{dx}(y - 3)^2 = \frac{d}{dx}(5)$$

Step 3: Use the chain rule to differentiate $(y - 3)^2$ with respect to x.

$$2(x - 2) + \frac{d}{dy}(y - 3)^2\frac{dy}{dx} = 0$$

$$2x - 4 + 2(y - 3)\frac{dy}{dx} = 0$$

Step 4: Substitute the given point and rearrange to find the gradient of the tangent.

At (1, 5),

$$2 - 4 + 2(5 - 3)\frac{dy}{dx} = 0$$

$$4\frac{dy}{dx} = 2$$

$$\frac{dy}{dx} = \tfrac{1}{2}$$

Tip:
It's easier to substitute in and then rearrange. As $\frac{dy}{dx}$ is in terms of x and y, you'll need to substitute both coordinates of the point.

Step 5: Find the gradient of the normal at the given point.

The gradient of the normal at (1, 5) is -2

An equation of the normal to the circle at (1, 5) is

Step 6: Use an appropriate straight line equation.

$$y - 5 = -2(x - 1)$$
$$y = 7 - 2x$$

Recall:
Tangents and normals are perpendicular, so the product of their gradients is -1
(C1 Section 2.2).

Recall:
Equation of a straight line
(C1 Section 2.1).

Parametric functions

To differentiate a function defined parametrically in terms of t:

- differentiate x with respect to t
- differentiate y with respect to t
- use a version of the chain rule, $\frac{dy}{dx} = \frac{dy}{dt} \div \frac{dx}{dt}$.

By the chain rule,

$$\frac{dy}{dx} = \frac{dy}{dt} \times \frac{dt}{dx}$$

As $\frac{dt}{dx} = \frac{1}{\frac{dx}{dt}}$, dividing by $\frac{dx}{dt}$ is the same as multiplying by $\frac{dt}{dx}$.

So

$$\frac{dy}{dx} = \frac{dy}{dt} \div \frac{dx}{dt}$$

Example 4.3 Find an equation of the tangent to the curve with parametric equations $x = 1 - t^3$, $y = 3t^2$, at the point where $t = -1$.

Step 1: Differentiate x and y with respect to t.

$x = 1 - t^3 \qquad y = 3t^2$

$\frac{dx}{dt} = -3t^2 \qquad \frac{dy}{dt} = 6t$

Step 2: Use the chain rule to find $\frac{dy}{dx}$.

$\frac{dy}{dx} = \frac{dy}{dt} \div \frac{dx}{dt} = -\frac{6t}{3t^2} = -\frac{2}{t}$

Step 3: Find $\frac{dy}{dx}$ at the given point.

When $t = -1$,

$$\frac{dy}{dx} = -\frac{2}{t} = -\frac{2}{-1} = 2$$

The gradient of the tangent at $t = -1$ is 2.

Step 4: Substitute the value of t to find the x- and y-coordinates.

When $t = -1$,

$x = 1 - t^3 = 1 - (-1)^3 = 2$

$y = 3t^2 = 3 \times (-1)^2 = 3$

Step 5: Use an appropriate straight line equation.

An equation of the tangent at $t = -1$ is

$$y - 3 = 2(x - 2)$$
$$y - 3 = 2x - 4$$
$$y = 2x - 1$$

Note:
You could find the Cartesian equation and differentiate that, but the working is more complex.

Tip:
You could have found the reciprocal of $\frac{dx}{dt}$ and used $\frac{dy}{dx} = \frac{dy}{dt} \times \frac{dt}{dx}$, but the new format of the chain rule is easier to apply here.

Recall:
Finding the equation of a tangent (C1 Section 4.4).

Example 4.4 A curve has parametric equations $x = 3 \sin \theta + 2$, $y = \cos 2\theta + 4$, $-\frac{\pi}{2} \leqslant \theta \leqslant \frac{\pi}{2}$.

Find the coordinates of the stationary point on the curve.

Step 1: Differentiate x and y with respect to θ.

$x = 3 \sin \theta + 2 \qquad y = \cos 2\theta + 4$

$\frac{dx}{d\theta} = 3 \cos \theta \qquad \frac{dy}{d\theta} = -2 \sin 2\theta$

Recall:
$\frac{d}{d\theta}(\cos k\theta) = -k \sin k\theta$
(C3 Section 4.1).

Step 2: Use the chain rule to find $\frac{dy}{dx}$.

$$\frac{dy}{dx} = \frac{dy}{dt} \div \frac{dx}{dt} = -\frac{2\sin 2\theta}{3\cos\theta}$$
$$= -\frac{2 \times 2\sin\theta\cos\theta}{3\cos\theta}$$
$$= -\frac{4\sin\theta}{3}$$

Recall:
$\sin 2\theta \equiv 2\sin\theta\cos\theta$ (C3 Section 2.3).

Step 3: Set $\frac{dy}{dx} = 0$ and solve for θ.

At the stationary point,

$$\frac{dy}{dx} = 0 \Rightarrow -\frac{4\sin\theta}{3} = 0$$
$$\sin\theta = 0$$
$$\theta = 0$$

Recall:
Stationary points have zero gradient (C2 Section 6.1).

Tip:
Here there's only one value of θ in the range, but don't forget to solve for all possible values.

Step 4: Substitute the value of θ to find the x- and y-coordinates.

When $\theta = 0$,

$$x = 3\sin\theta + 2 = 3\sin 0 + 2 = 2$$
$$y = \cos 2\theta + 4 = \cos 0 + 4 = 5$$

There is a stationary point at (2, 5).

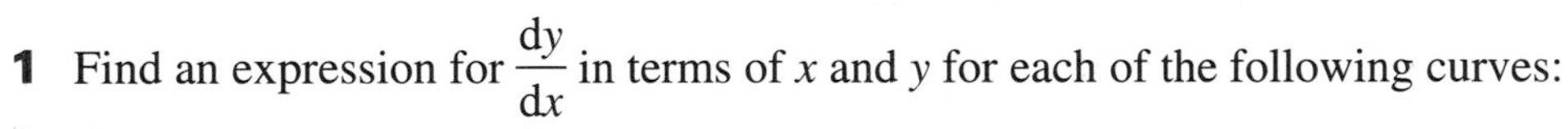

SKILLS CHECK 4A: Differentiation of implicit and parametric functions

1 Find an expression for $\frac{dy}{dx}$ in terms of x and y for each of the following curves:

a $3x^2 - y^2 + 5x - 6y + 5 = 0$

b $y^3 + x^2y - 2x = 0$

c $y^3 + x\ln y = 3x^2$

2 Find the coordinates of the stationary points on the curve $x^2 + y^2 - 6x - 8y + 24 = 0$.

3 A curve has equation $x^2 + 3y^2 - 4x - 6y - 14 = 0$.

a Use implicit differentiation to find $\frac{dy}{dx}$.

b Find an equation of the normal to the curve at the point $(5, -1)$.

4 The curves $xy = 2$ and $y^2 - x^2 = 3$ intersect at the point (1, 2).

a Find the gradients of the tangents to the curves at the point of intersection.

b Determine the angle between these tangents.

5 Find an expression for $\frac{dy}{dx}$ in terms of t for each of the following curves:

a $x = 2t^3,\ y = 3t^2 + 1$ **b** $x = 10\sec t,\ y = 5\tan t$ **c** $x = t^2 - 4t,\ y = t\ln 4t$

6 A curve has parametric equations $x = t^2 + 1,\ y = 4(t + 1)$.

a Find $\frac{dy}{dx}$ in terms of t.

The normal to the curve at (2, 8) cuts the x-axis at $(p, 0)$ and the y-axis at $(0, q)$.

b Find an equation of the normal at (2, 8).

c Find p and q.

7 A curve has parametric equations $x = 6t + 3\sin 2t$, $y = 2\cos 2t$, $0 < t < \frac{\pi}{2}$.
Show that $\frac{dy}{dx} = -\frac{2}{3}\tan t$.

 **8** A curve has parametric equations $x = at$, $y = at^2 + at$, where a is a constant. Given that the curve has a stationary point at $(-2, -1)$, find a.

9 A curve has parametric equations $x = e^{2t} + 2t$, $y = e^t - t$.
Find the coordinates of the stationary point on the curve.

SKILLS CHECK **4A EXTRA is on the CD**

4.2 Exponential growth and decay

Exponential growth and decay.

In *Core 3* you studied simple exponential growth and decay models using the exponential function $f(t) = e^t$.

The function $f(t) = a^t$, where a is a constant, can also be used for such modelling.

The derivatives of these functions are used to find the rate of growth or decay at a moment in time.

You already know that $\frac{d}{dx}(e^x) = e^x$. In *Core 4* you need to be able to find $\frac{d}{dx}(a^x)$. This is illustrated in the following example.

Recall: When $k > 0$, $y = Ae^{kt}$ can be used to model exponential growth and $y = Ae^{-kt}$ can be used to model exponential decay (C3 Section 3.2).

Example 4.5 Given that $y = a^x$, where a is a positive constant, find $\frac{dy}{dx}$.

Step 1: Take natural logs of both sides.

$$y = a^x$$
$$\ln y = \ln a^x$$
$$= x\ln a$$

Recall: $\log a^n = n \log a$ (C2 Section 5.2).

Step 2: Differentiate term by term with respect to x.

Differentiating with respect to x,

$$\frac{d}{dx}(\ln y) = \frac{d}{dx}(x\ln a)$$

Tip: $\ln a$ is a constant.

Step 3: Use the chain rule to differentiate $\ln y$ with respect to x.

$$\frac{d}{dy}(\ln y)\frac{dy}{dx} = \ln a$$

$$\frac{1}{y}\frac{dy}{dx} = \ln a$$

Recall: The derivative of $\ln y$ is $\frac{1}{y}$ (C3 Section 4.1).

Tip: Multiply both sides by y.

Step 4: Rearrange to make $\frac{dy}{dx}$ the subject.

$$\frac{dy}{dx} = y\ln a$$

Step 5: Substitute in for y.

$$= a^x \ln a$$

So

- if $y = a^x$, $\frac{dy}{dx} = a^x \ln a$
- if $y = e^x$, $\frac{dy}{dx} = e^x \ln e = e^x$

Tip: You need to remember this result.

Note: This proves the result that you learned in C3.

Example 4.6 A curve has equation $y = 3^x + 3^{-x} + \frac{2}{3}$.
Find an equation of the tangent to the curve at the point where $x = 1$.

Step 1: Differentiate term by term.

$$y = 3^x + 3^{-x} + \tfrac{2}{3}$$

$$\frac{dy}{dx} = 3^x \ln 3 - 3^{-x} \ln 3$$

Step 2: Find $\frac{dy}{dx}$ at the given point.

When $x = 1$,

$$\frac{dy}{dx} = 3^1 \ln 3 - 3^{-1} \ln 3$$

$$= 3 \ln 3 - \tfrac{1}{3} \ln 3 = \tfrac{8}{3} \ln 3$$

Step 3: Find the y-coordinate at the given point.

When $x = 1$,

$$y = 3^1 + 3^{-1} + \tfrac{2}{3} = 3 + \tfrac{1}{3} + \tfrac{2}{3} = 4$$

Step 4: Use an appropriate straight line equation.

An equation of the tangent at (1, 4) is

$$y - 4 = \tfrac{8}{3} \ln 3(x - 1)$$

$$y = \tfrac{8}{3} \ln 3(x - 1) + 4$$

Tip:
Use the chain rule to differentiate 3^{-x} with $t = -x$:

$$\frac{d}{dx}(3^{-x})$$
$$= \frac{d}{dt}(3^t) \times \frac{d}{dx}(-x)$$
$$= (3^t \ln 3) \times -1$$
$$= -3^{-x} \ln 3$$

Tip:
There is no need to simplify the equation further, unless the question asks for a specific format.

Example 4.7 A motorbike cost £8000 when it was new at the end of 2004. By the end of 2006 it has an estimated value of £5120.

The value of the motorbike can be modelled by an equation of the form

$$V = V_0(k)^{-t}$$

where £V is the value of the motorbike at time t years from 2004 and k is a positive constant.

a Find the value of k.

b Estimate the value of $\frac{dV}{dt}$ at the end of 2008.

c What does this value represent?

Step 1: Substitute the given values into the equation.

a When $t = 2$, $V = 5120$ and $V_0 = 8000$.

So

$$5120 = 8000(k)^{-2}$$

Step 2: Rearrange to find k.

$$\frac{5120}{8000} = \frac{1}{k^2}$$

$$k^2 = \frac{8000}{5120} = 1.5625$$

$$k = \sqrt{1.5625} = 1.25$$

Step 3: Differentiate V with respect to t using the result for $\frac{d}{dx}(a^x)$.

b $V = 8000(1.25)^{-t}$

$$\frac{dV}{dt} = 8000(-1.25^{-t} \ln 1.25)$$

At the end of 2008, $t = 4$, so

Step 4: Find $\frac{dV}{dt}$ for the given value of t.

$$\frac{dV}{dt} = 8000(-1.25^{-4} \ln 1.25) = -3276.8 \ln 1.25$$

$$= -731.19\ldots$$

$$= -731 \text{ (3 s.f.)}$$

Tip:
V_0 is the value of the motorbike when $t = 0$, i.e. the initial cost.

Tip:
In 2006, $t = 2006 - 2004 = 2$.

Tip:
As k is positive you only need the positive value of the square root.

Tip:
Use the chain rule to differentiate $(1.25)^{-t}$ as in the above example.

Tip:
$t = 2008 - 2004 = 4$

Tip:
Don't simplify until the end to avoid rounding errors.

Step 5: Interpret your answer.

c This is the rate of increase of the motorbike's value during 2008. As the value is negative, the motorbike is depreciating in value at a rate of £731 per year.

Note: Depreciation is a loss of value.

Interpreting graphs

In the above example, sketching V against t, where $V = 8000(1.25)^{-t}$ would give:

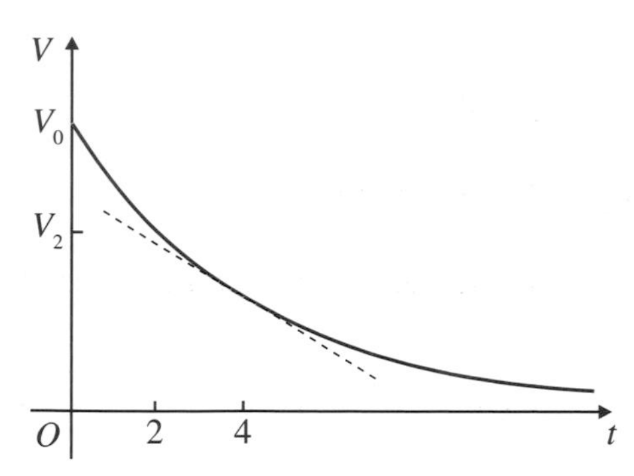

The intersection with the V-axis gives the initial value of the motorbike (i.e. £8000).

In 2006, $t = 2$, $V = V_2 = 5120$.

The gradient of the graph at $t = 4$ is the rate of change in value when the motorbike is four years old in 2008.

The graph is much steeper at the start so the motorbike loses value faster when it is nearly new.
Eventually the motorbike will be worth almost nothing although, using this model, its value will never be zero.

4.3 Formation of differential equations

Formation of simple differential equations.

Connected rates of change

The chain rule can be applied to **connected rates of change**. This is illustrated in the following example.

Example 4.8 A cuboid has a rectangular base of length $4x$ cm and width x cm. The height of the cuboid is $3x$ cm.

a Given that the width of the base is increasing at a rate of 3 cm/s, find the rate of increase of the area of the base when $x = 2$.

b Given that the volume of the cuboid is increasing at a rate of 18 cm³/s, find the rate at which the width of the base is increasing when $x = 1$.

Step 1: Define the variables, stating the units.
Step 2: Express the given rate of change in calculus form.
Step 3: Express A in terms of x and differentiate with respect to x.

a Let A be the area of the base, in cm², and t the time in seconds.

It is given that $\dfrac{dx}{dt} = 3$.

$A = 4x^2 \Rightarrow \dfrac{dA}{dx} = 8x$

Tip: The rate of change of the width x, with time, is given by $\dfrac{dx}{dt}$.

Step 4: Use the chain rule to differentiate A with respect to t.

We require the rate of change of A, i.e. $\frac{dA}{dt}$.

By the chain rule,

$$\frac{dA}{dt} = \frac{dA}{dx} \times \frac{dx}{dt}$$

Step 5: Substitute known values.

$$= 8x \times 3$$
$$= 24x$$

When $x = 2$, $\frac{dA}{dt} = 24 \times 2 = 48$

The area is increasing at 48 cm^2/s when $x = 2$.

Tip: The rate at which the area is changing (with time) is given by $\frac{dA}{dt}$.

Tip: Specify the units in your final answer.

Step 1: Define the variables, stating the units.

b Let V be the volume of the cuboid, in cm^3, and t the time in seconds.

Step 2: Express the given rate of change in calculus form.

Volume increasing at 18 $cm^3/s \Rightarrow \frac{dV}{dt} = 18$

Step 3: Express V in terms of x and differentiate with respect to x.

$V = 12x^3 \Rightarrow \frac{dV}{dx} = 36x^2$

Tip: The rate at which the volume is changing (with time) is given by $\frac{dV}{dt}$.

Step 4: Use the chain rule to differentiate x with respect to t.

We require the rate of change of x, i.e. $\frac{dx}{dt}$.

By the chain rule,

$$\frac{dx}{dt} = \frac{dx}{dV} \times \frac{dV}{dt}$$

Since $\frac{dV}{dx} = 36x^2$, $\frac{dx}{dV} = \frac{1}{36x^2}$

Recall: $\frac{dx}{dV} = \frac{1}{\frac{dV}{dx}}$ (C3 Section 4.3).

Step 5: Substitute known values.

So $\frac{dx}{dt} = \frac{1}{36x^2} \times 18$

When $x = 1$, $\frac{dx}{dt} = \frac{1}{36 \times 1^2} \times 18 = 0.5$

The width of the base is increasing at 0.5 cm/s when $x = 1$.

Example 4.9 The variables P and Q are related by the expression $P = 5Q^2 + 3$. If Q is decreasing at a rate of 4 units/s, find the rate at which P is decreasing when $Q = 7$.

Step 1: Express the given rate of change in calculus form.

Q is decreasing at a rate of 4 units/s $\Rightarrow \frac{dQ}{dt} = -4$

Tip: You need the negative, since Q is decreasing.

Step 2: Differentiate P with respect to Q.

$P = 5Q^2 + 3 \Rightarrow \frac{dP}{dQ} = 10Q$

Step 3: Use the chain rule to differentiate P with respect to t.

$$\frac{dP}{dt} = \frac{dP}{dQ} \times \frac{dQ}{dt}$$
$$= 10Q \times (-4)$$
$$= -40Q$$

Step 4: Substitute known values.

When $Q = 7$, $\frac{dP}{dt} = -40 \times 7 = -280$

So P is decreasing at a rate of 280 units/s.

Note: Since $\frac{dP}{dt} < 0$, P is also decreasing.

Simple differential equations

A differential equation is an equation that contains a gradient function, for example:

$$\frac{dy}{dx} = 2x, \quad y\frac{dy}{dx} = e^x + 2, \quad \frac{d^2y}{dx^2} = -12x^2, \quad \frac{dV}{dt} = \tfrac{1}{2}V.$$

This chapter deals with *forming* differential equations; *solving* them is covered in Chapter 5.

Differential equations are often used to model economic, social or scientific situations. Many scientific laws can be expressed in terms of a differential equation.

Note: In C4 you study first order differential equations, i.e. equations containing the first derivative only.

Note: Solving differential equations (see Section 5.6).

Example 4.10 A mathematical model for the number of bacteria, N, in an experiment states that N is increasing at a rate proportional to the number of bacteria present at time t.

Write down a differential equation involving N and t.

Step 1: Use the given information to form a differential equation.

The rate of change of N is $\frac{dN}{dt}$.

$$\frac{dN}{dt} \propto N$$

i.e. $\frac{dN}{dt} = kN$, where k is a positive constant

Note: The symbol $\propto$ means 'is proportional to'.

Note: In the exam you would then be asked to solve the equation. (see Section 5.6).

Example 4.11 According to Newton's law of cooling, the rate of temperature loss of a body is proportional to the difference between the temperature of the body and the temperature of the air.

Write a differential equation for this law.

Step 1: Define the variables.

Let T be the temperature of the body at time t seconds, and let T_0 be the air temperature.

Step 2: Use the given information to form a differential equation.

The rate of change of T is $\frac{dT}{dt}$.

$$\frac{dT}{dt} \propto -(T - T_0)$$

i.e. $\frac{dT}{dt} = -k(T - T_0)$, where $k > 0$

Tip: Use a minus sign to show that the temperature is decreasing.

Tip: $T - T_0$ is the difference between the temperature of the body and the air temperature.

Example 4.12 At time t, the surface area of a circular oil slick with radius r is increasing at a constant rate of $50\ \text{m}^2/\text{s}$.

Find a differential equation for the rate of change of the radius in terms of r and t.

Step 1: Define the variables.

Let the surface area of the oil slick at time t seconds be $A\ \text{m}^2$.

Step 2: Express the given rate of change in calculus form.

A is increasing at a rate of $50\ \text{m}^2/\text{s} \Rightarrow \frac{dA}{dt} = 50$

Step 3: Express A in terms of r and differentiate with respect to r.

$A = \pi r^2 \Rightarrow \frac{dA}{dr} = 2\pi r$

We require the rate of change of r, i.e. $\frac{dr}{dt}$.

Tip: You need to use connected rates of change to set up this differential equation.

Step 4: Use the chain rule to differentiate r with respect to t.

By the chain rule,

$$\frac{dr}{dt} = \frac{dr}{dA} \times \frac{dA}{dt}$$

Since $\frac{dA}{dr} = 2\pi r$, $\frac{dr}{dA} = \frac{1}{2\pi r}$

Step 5: Substitute known values.

So $\frac{dr}{dt} = \frac{1}{2\pi r} \times 50$

$$\frac{dr}{dt} = \frac{25}{\pi r}$$

Recall:

$\frac{dr}{dA} = \frac{1}{\frac{dA}{dr}}$ (C3 Section 4.3).

SKILLS CHECK 4B: Exponential growth and decay, and formation of differential equations

1 Differentiate the following:

a $y = 5^x$

b $y = \left(\frac{1}{2}\right)^t$

c $y = x^2 a^x$

d $y = 3^{1 + 2t}$

2 A curve has parametric equations $x = \sin t$, $y = 2^t$, $-\frac{1}{2}\pi < t < \frac{1}{2}\pi$.

a Find $\frac{dy}{dx}$ in terms of t.

b Find the equation of the tangent to the curve at (0, 1), giving your answer in the form $y = x \ln a + b$, where a and b are integers.

3 Scientists are monitoring the decay of a radioactive isotope. Initially there are 1.25×10^{17} atoms of the radioactive isotope. After two days, there are 8×10^{16} radioactive atoms present.

An equation of the form $N = N_0(k)^{-t}$ is used to model the data, where N is the number of radioactive atoms present after t days, N_0 is the initial number of atoms recorded and k is a positive constant.

a Find k.

b Find the value of $\frac{dN}{dt}$ after 10 days, giving your answer to three significant figures.

c Explain the significance of your answer to **b**.

4 Variables x and y are connected by the equation $y = 3x^2 + 5 \ln x$.

a Given that x is increasing at the rate of 4 units per second, find the rate of increase in y when $x = 2$.

b Given that x is decreasing at the rate of 4 units per second, find the rate at which y is changing when $x = 3$, stating whether this is an increase or a decrease.

5 The radius of a circular ink blot is increasing at a rate of 0.3 cm/s. Find the rate at which the area is increasing when the radius is 0.5 cm.

6 The volume of a spherical balloon is decreasing at a rate of 25 cm^3/s. Find the rate at which the radius is decreasing at the instant when the radius of the balloon is 2 cm. Give your answer to two significant figures.

7 The gradient of a curve C at any point is proportional to the sum of the Cartesian coordinates of that point. Given that, at the point $(0, -3)$ on C, the gradient is $-\frac{3}{4}$, form a differential equation in terms of x and y.

8 The volume of air inside a spherical football placed in the sun expands at a rate proportional to the surface area of the ball.

The volume of the ball, with radius r, is $\frac{4}{3}\pi r^3$ and the surface area is $4\pi r^2$. Form a differential equation to model the rate of change in the radius of the ball.

9 A mathematical model of an ice cube melting assumes that the surface area is decreasing at a constant rate of 0.2 cm²/s. Form a differential equation for the rate of change of the volume, V, of the ice cube at time t seconds, giving your answer in terms of V and t.

SKILLS CHECK **4B EXTRA is on the CD**

Examination practice 4: Differentiation

1 Given that $e^{2x} + e^{2y} = xy$, find $\frac{dy}{dx}$ in terms of x and y. [London June 1999]

2 **a** Given that $x = 2^t$, by using logarithms, prove that

$$\frac{dx}{dt} = 2^t \ln 2.$$

A curve C has parametric equations

$$x = 2^t, \quad y = 3t^2.$$

The tangent to C at the point with coordinates $(2, 3)$ cuts the x-axis at the point P.

b Find $\frac{dy}{dx}$ in terms of t.

c Calculate the x-coordinate of P, giving your answer to 3 decimal places. [Edexcel June 2000]

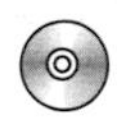

3

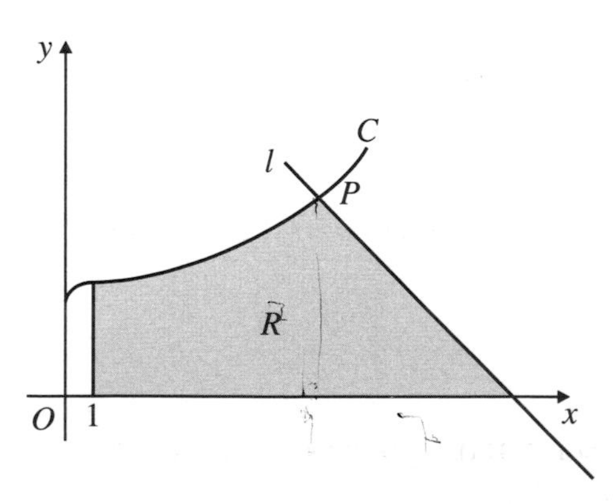

The diagram shows part of the curve C with parametric equations

$$x = (t + 1)^2, \quad y = \tfrac{1}{2}t^3 + 3, \quad t \geq -1.$$

P is the point on the curve where $t = 2$. The line l is the normal to C at P.

a Find an equation of l.

The shaded region R is bounded by C, l, the x-axis and the line with equation $x = 1$.

b Using integration and showing all your working, find the area of R. [London Jan 2000]

4 A curve, C, is given by

$$x = 2t + 3, \quad y = t^3 - 4t,$$

where t is a parameter. The point A has parameter $t = -1$ and the line l is the tangent to C at A. The line l also intersects the curve at B.

a Show that an equation for l is $2y + x = 7$.

b Find the value of t at B. [London June 1998]

5 The curve C has parametric equations

$$x = a \sec t, \quad y = b \tan t, \quad 0 < t < \frac{\pi}{2},$$

where a and b are positive constants.

a Prove that $\frac{dy}{dx} = \frac{b}{a} \text{cosec } t$.

b Find the equation in the form $y = px + q$ of the tangent to C at the point where $t = \frac{\pi}{4}$.

[Edexcel P3 Mock]

6 A curve has equation $7x^2 + 48xy - 7y^2 + 75 = 0$.

A and B are two distinct points on the curve. At each of these points the gradient of the curve is equal to $\frac{2}{11}$.

a Use implicit differentiation to show that $x + 2y = 0$ at the points A and B.

b Find the coordinates of the points A and B. [Edexcel June 2003]

7 The circle C has equation $x^2 + y^2 - 8x - 16y - 209 = 0$.

a Find the coordinates of the centre of C and the radius of C.

The point $P(x, y)$ lies on C.

b Find, in terms of x and y, the gradient of the tangent to C at P.

c Hence or otherwise, find an equation of the tangent to C at the point $(21, 8)$.

[Edexcel May 2002]

8 The curve C has equation

$$13x^2 + 13y^2 - 10xy = 52.$$

Find an expression for $\frac{dy}{dx}$ as a function of x and y, simplifying your answer. [Edexcel C4 Specimen]

9 A curve has equation

$$x^3 - 2xy - 4x + y^3 - 51 = 0.$$

Find an equation of the normal to the curve at the point $(4, 3)$, giving your answer in the form $ax + by + c = 0$, where a, b and c are integers. [Edexcel C4 Mock]

10 The value £V of a car t years after the 1st January 2001 is given by the formula

$$V = 10\,000 \times (1.5)^{-t}.$$

a Find the value of the car on 1st January 2005.

b Find the value of $\frac{dV}{dt}$ when $t = 4$.

c Explain what the answer to part **b** represents. [Edexcel May 2005]

 **11** The filter paper in a coffee machine is an inverted cone with the angle at the vertex measuring 120°. Coffee drips out at a rate of 2.5 cm^3/s.

a Show that the volume of coffee in the filter paper is given by πh^3, where h is the vertical depth of the coffee at time t seconds.

b Show that $\frac{dh}{dt} = \frac{k}{h^2}$, where k is a constant to be found.

5 Integration

5.1 Integration of standard functions

Integrate e^x, $\frac{1}{x}$, $\sin x$, $\cos x$.

Recall: Integration (C1 Chapter 5, C2 Chapter 7).

In *Core 3* you differentiated functions such as e^{kx}, $\ln kx$, $\sin kx$, $\cos kx$ and $\tan kx$, with the aid of the chain rule.

Recall: Differentiation (C3 Section 4.1).

Reversing this process leads to the following **standard integrals**, which should be learnt for *Core 4*.

$\mathbf{f(x)}$	$\int \mathbf{f(x)\,dx}$
e^{kx}	$\frac{1}{k}e^{kx} + c$
$\frac{1}{x}$	$\ln\lvert x\rvert + c$
$\cos kx$	$\frac{1}{k}\sin kx + c$
$\sin kx$	$-\frac{1}{k}\cos kx + c$
$\sec^2 kx$	$\frac{1}{k}\tan kx + c$

Note: It is possible to incorporate the constant into the log expression by letting $c = \ln k$. Then $\int \frac{1}{x}\,dx = \ln\lvert x\rvert + \ln k = \ln k\lvert x\rvert$.

Note: When integrating trig functions, radians must be used.

Note: $\int \sec^2 kx\,dx$ is given in the formula booklet.

Example 5.1 Find

a $\int (x^2 - \sin 3x + 2e^{5x})\,dx$

b $\int \left(6\sec^2 2x + \frac{1}{3x^2}\right) dx$

c i $\int \frac{2}{x}\,dx$ **ii** $\int \frac{1}{2x}\,dx$

Step 1: Integrate term by term.

a $\int (x^2 - \sin 3x + 2e^{5x})\,dx = \frac{1}{3}x^3 - (-\frac{1}{3}\cos 3x) + 2 \times \frac{1}{5}e^{5x} + c$

$= \frac{1}{3}x^3 + \frac{1}{3}\cos 3x + \frac{2}{5}e^{5x} + c$

Tip: Take care with fractions and signs.

Step 1: Write terms in index form, where appropriate, before integrating.

Step 2: Integrate term by term.

b $\int \left(6\sec^2 2x + \frac{1}{3x^2}\right) dx = \int (6\sec^2 2x + \frac{1}{3}x^{-2})\,dx$

$= 6 \times \frac{1}{2}\tan 2x + \frac{1}{3} \times \frac{x^{-1}}{-1} + c$

$= 3\tan 2x - \frac{1}{3x} + c$

Tip: Remember to include the integration constant.

Step 1: Take out a numerical factor then integrate using a standard function.

c i $\int \frac{2}{x}\,dx = 2\int \frac{1}{x}\,dx$

$= 2\ln\lvert x\rvert + c$

Tip: Take great care when the number is in the denominator.

ii $\int \frac{1}{2x}\,dx = \frac{1}{2}\int \frac{1}{x}\,dx$

$= \frac{1}{2}\ln\lvert x\rvert + c$

Example 5.2 Given that $\int_{-4}^{-2}\left(x+\frac{5}{x}\right)\mathrm{d}x = p + q\ln 2$, where p and q are integers, find the value of p and the value of q.

Step 1: Integrate term by term.

$$\int_{-4}^{-2}\left(x+\frac{5}{x}\right)\mathrm{d}x = \left[\tfrac{1}{2}x^2 + 5\ln|x|\right]_{-4}^{-2}$$

Step 2: Substitute the limits and evaluate.

$$= \tfrac{1}{2}\times(-2)^2 + 5\ln|-2| - (\tfrac{1}{2}\times(-4)^2 + 5\ln|-4|)$$
$$= 2 + 5\ln 2 - (8 + 5\ln 4)$$
$$= -6 + 5\ln\tfrac{2}{4}$$
$$= -6 + 5\ln 2^{-1}$$
$$= -6 - 5\ln 2$$

Step 3: Compare with the given result.

Hence $p = -6$ and $q = -5$

Recall: Definite integration (C2 Section 7.1).

Note: The modulus function is very important here.

Tip: If $a > 0$, $\ln|-a| = \ln a$

Recall: $\log a - \log b = \log\frac{a}{b}$ and $\log a^n = n\log a$ (C2 Section 5.2).

Example 5.3 The diagram shows a sketch of the curve $y = 2 - \mathrm{e}^x$. The curve crosses the x-axis at P. The region R, bounded by the curve and the coordinate axes, is shown shaded in the diagram.

Recall: Area under a curve (C2 Section 7.2).

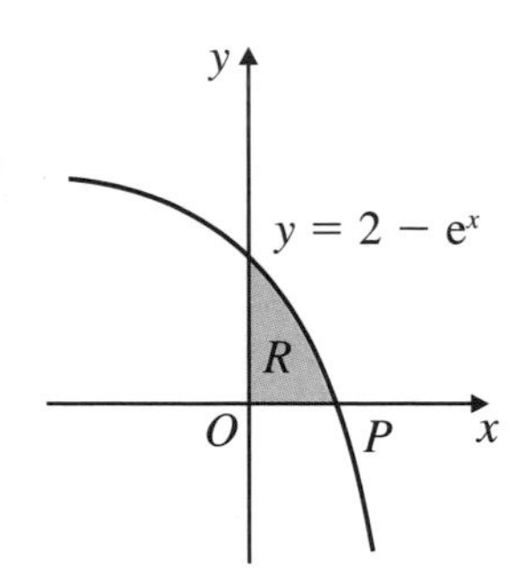

a Find the exact x-coordinate of P.

b The area of R is $\ln a + b$, where a and b are integers. Find the values of a and b.

Step 1: Set $y = 0$ and solve.

a At P, $y = 0$

$\Rightarrow$ $2 - \mathrm{e}^x = 0$

$\mathrm{e}^x = 2$

$x = \ln 2$

The x-coordinate of P is $\ln 2$.

Recall: $\mathrm{e}^a = b \Leftrightarrow a = \ln b$

Tip: Notice the instruction to give the exact value. Your calculator will only give an approximate value for ln 2.

Step 1: Integrate using the area formula.

b $\int_b^a y\,\mathrm{d}x = \int_0^{\ln 2}(2 - \mathrm{e}^x)\,\mathrm{d}x$

Step 2: Substitute the limits and evaluate.

$$= \left[2x - \mathrm{e}^x\right]_0^{\ln 2}$$
$$= 2\ln 2 - \mathrm{e}^{\ln 2} - (0 - \mathrm{e}^0)$$
$$= 2\ln 2 - 2 + 1$$
$$= \ln 4 - 1$$

Step 3: Write in the required form and compare values to find a and b.

Hence $a = 4$ and $b = -1$.

Recall: $\mathrm{e}^{\ln a} = a$, $\mathrm{e}^0 = 1$, $n\ln a = \ln a^n$

Tip: Do not assume when you substitute a limit of zero that the value(s) will be zero.

Integrals of the type $\int \frac{f'(x)}{f(x)}\,dx$

In *Core 3*, you applied the chain rule to differentiate log functions. Consider the general case, where $y = \ln f(x)$. Using the chain rule it can be shown that $\frac{dy}{dx} = \frac{f'(x)}{f(x)}$. Reversing this process gives the following result, which can be quoted in the examination:

$$\int \frac{f'(x)}{f(x)}\,dx = \ln|f(x)| + c$$

Tip: Look for a fractional expression where the numerator is a derivative, or a multiple of the derivative, of the denominator.

Example 5.4 Find:

a $\int \frac{1}{2+3x}\,dx$ **b** $\int \frac{\cos x}{1+2\sin x}\,dx$ **c** $\int \frac{e^{-x}}{1-e^{-x}}\,dx$

Step 1: Check whether the numerator is a multiple of the derivative of the denominator.

Step 2: If necessary adjust using a numerical factor.

a $\frac{d}{dx}(2+3x) = 3$

$$\int \frac{1}{2+3x}\,dx = \tfrac{1}{3}\ln|2+3x| + c$$

Tip: In **a**, the numerator is $\frac{1}{3}$ of the derivative of the denominator, so use $\frac{1}{3}$ as an adjustment factor.

b $\frac{d}{dx}(1+2\sin x) = 2\cos x$

$$\int \frac{\cos x}{1+2\sin x}\,dx = \tfrac{1}{2}\ln|1+2\sin x| + c$$

Tip: In **b**, the numerator is $\frac{1}{2}$ the derivative of the denominator, so use $\frac{1}{2}$ as an adjustment factor.

c $\frac{d}{dx}(1-e^{-x}) = e^{-x}$

$$\int \frac{e^{-x}}{1-e^{-x}}\,dx = \ln|1-e^{-x}| + c$$

Tip: In **c**, no adjustment factor is necessary.

Example 5.5 **a** By expressing $\tan x$ in terms of $\sin x$ and $\cos x$, show that

$$\int \tan x\,dx = \ln|\sec x| + c$$

b By expressing $\cot x$ in terms of $\sin x$ and $\cos x$, find $\int \cot x\,dx$.

Step 1: Use an appropriate trig identity.

Step 2: Integrate, recognising an appropriate standard result.

a $\tan x = \frac{\sin x}{\cos x} \Rightarrow \int \tan x\,dx = \int \frac{\sin x}{\cos x}\,dx$

$$= -\ln|\cos x| + c$$

$$= \ln|(\cos x)^{-1}| + c$$

$$= \ln|\sec x| + c$$

Tip: Since $\frac{d}{dx}(\cos x) = -\sin x$, spot the use of $\int \frac{f'(x)}{f(x)}\,dx = \ln|f(x)| + c$

Tip: $n\log x = \log x^n$

b $\cot x = \frac{\cos x}{\sin x} \Rightarrow \int \cot x\,dx = \int \frac{\cos x}{\sin x}\,dx$

$$= \ln|\sin x| + c$$

Tip: $\frac{d}{dx}(\sin x) = \cos x$, so the numerator is the derivative of the denominator.

Using identities to integrate trigonometric functions

The following identities, studied in *Core 2* and *Core 3*, are particularly useful for integrating some trigonometric functions:

$$\cos^2 A + \sin^2 A \equiv 1$$
$$1 + \tan^2 A \equiv \sec^2 A$$

Recall: Trigonometric identities (C3 Sections 2.2 and 2.3).

The double angle formulae:

$$\cos 2A \equiv 2\cos^2 A - 1 \Rightarrow \cos^2 A \equiv \tfrac{1}{2}(1 + \cos 2A)$$
$$\cos 2A \equiv 1 - 2\sin^2 A \Rightarrow \sin^2 A \equiv \tfrac{1}{2}(1 - \cos 2A)$$
$$\sin 2A \equiv 2\sin A\cos A$$

Note: The rearrangements with $\cos^2 A$ or $\sin^2 A$ as the subject are often used. It is helpful to learn them.

The addition formulae:

$$\sin(A \pm B) \equiv \sin A\cos B \pm \cos A\sin B$$
$$\cos(A \pm B) \equiv \cos A\cos B \mp \sin A\sin B$$

Example 5.6 Find:

a $\int \tan^2 x\,dx$ **b** $\int \cos^2 3x\,dx$

c $\int \sin 3x\cos 3x\,dx$ **d** $\int_0^{2\pi} \sin^2(\tfrac{1}{2}x)\,dx$

Step 1: Apply an appropriate trig identity.
Step 2: Integrate, term by term.

a $\int \tan^2 x\,dx = \int (\sec^2 x - 1)\,dx$
$= \tan x - x + c$

Tip: Rearrange $1 + \tan^2 A \equiv \sec^2 A$.

Recall: $\frac{d}{dx}(\tan x) = \sec^2 x$

Step 1: Apply an appropriate trig identity.
Step 2: Integrate, term by term.

b $\int \cos^2 3x\,dx = \int \tfrac{1}{2}(1 + \cos 6x)\,dx$
$= \tfrac{1}{2}\int (1 + \cos 6x)\,dx$
$= \tfrac{1}{2}(x + \tfrac{1}{6}\sin 6x) + c$

Tip: Rearrange $\cos 2A \equiv 2\cos^2 A - 1$.

Tip: Take out a numerical factor before integrating.

Step 1: Apply an appropriate trig identity.
Step 2: Integrate.

c $\int \sin 3x\cos 3x\,dx = \tfrac{1}{2}\int \sin 6x\,dx$
$= \tfrac{1}{2}(-\tfrac{1}{6}\cos 6x) + c$
$= -\tfrac{1}{12}\cos 6x + c$

Tip: Use the double angle formula for $\sin 2A$.

Step 1: Apply an appropriate trig identity.
Step 2: Integrate, term by term.
Step 3: Substitute the limits and evaluate.

d $\int_0^{2\pi} \sin^2(\tfrac{1}{2}x)\,dx = \int_0^{2\pi} \tfrac{1}{2}(1 - \cos x)\,dx$
$= \tfrac{1}{2}\Big[x - \sin x\Big]_0^{2\pi}$
$= \tfrac{1}{2}(2\pi - \sin 2\pi - (0 - \sin 0))$
$= \tfrac{1}{2}(2\pi - 0 - (0 - 0))$
$= \pi$

Tip: Rearrange $\cos 2A \equiv 1 - 2\sin^2 A$.

Recall: $\frac{d}{dx}(\sin x) = \cos x$

Example 5.7 The diagram shows a sketch of the ellipse defined by the parametric equations $x = 2\sin t,\ y = \cos t,\ 0 \leqslant t \leqslant 2\pi$.

Use integration to show that the area of the ellipse is 2π.

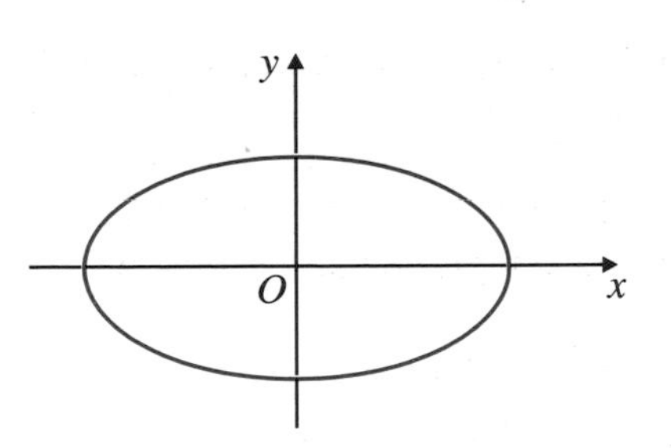

Step 1: Apply the area formula when the equation is given in terms of a parameter.

$$\text{Area} = \int_{t_1}^{t_2} y\frac{dx}{dt}\,dt$$

$$= \int_0^{2\pi} y\frac{dx}{dt}\,dt$$

Recall: Area under a curve given parametrically (Section 2.1).

Step 2: Find $\frac{dx}{dt}$.

$$x = 2\sin t \Rightarrow \frac{dx}{dt} = 2\cos t$$

Step 3: Substitute into the formula and integrate.

$$\int_0^{2\pi} y\frac{dx}{dt}\,dt = \int_0^{2\pi} \cos t \times 2\cos t\,dt$$

Step 4: Substitute the limits and evaluate.

$$= \int_0^{2\pi} 2\cos^2 t\,dt$$

$$= \int_0^{2\pi} (1 + \cos 2t)\,dt$$

$$= \left[t + \tfrac{1}{2}\sin 2t\right]_0^{2\pi}$$

$$= 2\pi + \tfrac{1}{2}\sin 4\pi - 0$$

$$= 2\pi$$

Tip: Rearrange $\cos 2A \equiv 2\cos^2 A - 1$.

The area of the ellipse is 2π.

Example 5.8 **a** Use the identities for $\sin(A + B)$ and $\sin(A - B)$ to show that

$$\sin(A + B) + \sin(A - B) \equiv 2\sin A\cos B$$

b Hence find $\int \sin 5x\cos 3x\,dx$.

Step 1: State the suggested trig identities and add.

a $\sin(A + B) \equiv \sin A\cos B + \cos A\sin B$

$\sin(A - B) \equiv \sin A\cos B - \cos A\sin B$

Hence $\sin(A + B) + \sin(A - B)$

$$\equiv \sin A\cos B + \cos A\sin B + \sin A\cos B - \cos A\sin B$$

$$\equiv 2\sin A\cos B$$

Note: These addition formulae are given in the formula booklet. Make sure that you transfer them accurately.

Step 1: Use the identity from part **a** to write **b** in different form.

b Let $A = 5x$ and $B = 3x$, then

$$2\sin 5x\cos 3x \equiv \sin(5x + 3x) + \sin(5x - 3x)$$

$$\equiv \sin 8x + \sin 2x$$

Hence

$$\sin 5x\cos 3x \equiv \tfrac{1}{2}(\sin 8x + \sin 2x)$$

Step 2: Integrate the expression term by term.

$$\int \sin 5x\cos 3x\,dx = \tfrac{1}{2}\int(\sin 8x + \sin 2x)\,dx$$

$$= \tfrac{1}{2}\left(\left(-\tfrac{1}{8}\cos 8x\right) - \tfrac{1}{2}\cos 2x\right) + c$$

$$= -\tfrac{1}{16}\cos 8x - \tfrac{1}{4}\cos 2x + c$$

Tip: Take out the numerical factor before integrating.

SKILLS CHECK 5A: Integration of standard functions

1 Find:

a $\int e^{3x+1}\,dx$ **b** $\int -e^{-u}\,du$ **c** $\int \frac{1}{e^{2t}}\,dt$

2 Find:

a $\int \frac{1}{3x}\,dx$ **b** $\int \frac{1}{1+5x}\,dx$ **c** $\int \frac{1}{3(2-x)}\,dx$

3 **a** Differentiate $e^{2x} + x^2$ with respect to x.

b Hence find $\int \frac{e^{2x}+x}{e^{2x}+x^2}\,dx$.

4 Find:

a $\int \cos \frac{1}{3}x\,dx$ **b** $\int \frac{\sin 2y}{2}\,dy$ **c** $\int \frac{3}{\cos^2 x}\,dx$

5 **a** Evaluate $\int_e^{e^2} \frac{1}{2x}\,dx$.

b Given that $\int_0^2 \frac{8x}{1+2x^2}\,dx = \ln a$, where a is an integer, find the value of a.

6 **a** Given that $\int_0^{\frac{1}{8}\pi} \tan 2x\,dx = a \ln 2$, where a is a rational number, find the value of a.

b Evaluate $\int_0^{\frac{1}{8}\pi} \tan^2 2x\,dx$.

7 Find the area enclosed by the curve $y = \sin x$, where $0 \leqslant x \leqslant \pi$, and the x-axis.

8 The gradient at the point (x, y) on the curve $y = f(x)$ is given by $e^{3x} - 2x$.
The curve goes through $(0, 1)$. Find an equation of the curve.

9 **a** Write $\cos^2 A$ in terms of $\cos 2A$.

b The diagram shows the graph of $y = \cos^2 x$ for $-\frac{1}{2}\pi \leqslant x \leqslant \frac{1}{2}\pi$.
Find the area enclosed by the curve and the x-axis.

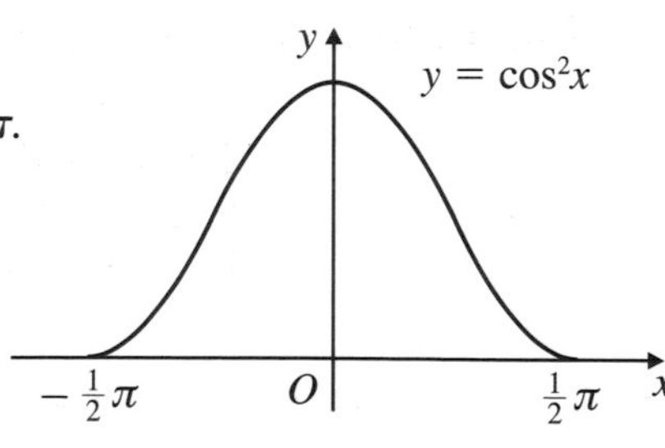

10 **a** Use the identities for $\cos(A+B)$ and $\cos(A-B)$ to show that

$$\cos(A+B) + \cos(A-B) \equiv 2\cos A \cos B$$

b Hence find $\int \cos 6x \cos 3x\,dx$.

11 Given that $\int_1^2 \left(\frac{2}{x} - 3x^2\right) dx = a + b\ln 2$, where a and b are integers, find the values of a and b.

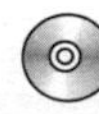

12 **a** Express $\cot x$ in terms of $\sin x$ and $\cos x$.

b By differentiating, show that $\frac{d}{dx}(\cot x) = -\operatorname{cosec}^2 x$.

c By using an appropriate trigonometric identity and your answer to part **b**, find $\int \cot^2 x\,dx$.

13 a Expand the brackets in the expression $(e^x - 1)(e^{2x} - 1)$.

b Hence find $\int_0^{0.2} (e^x - 1)(e^{2x} - 1)\,dx$, giving your answer to two significant figures.

SKILLS CHECK **5A EXTRA is on the CD**

5.2 Volume of revolution

Evaluation of volume of revolution.

When the region enclosed by a curve $y = f(x)$, the x-axis and the lines $x = a$ and $x = b$ is rotated completely about the x-axis, a solid is formed.

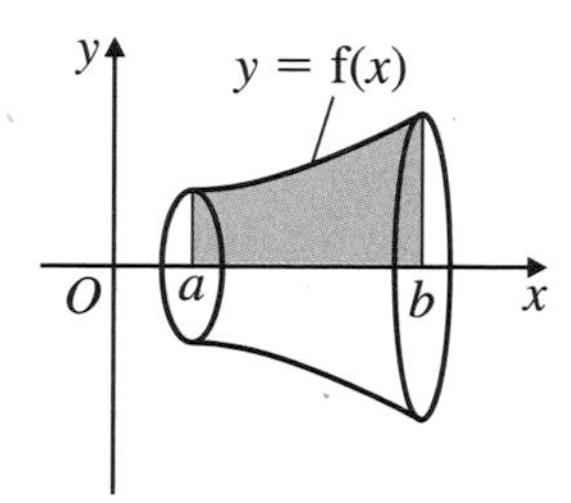

The volume of this solid is known as the **volume of revolution** and is evaluated using the formula

$$V = \pi \int_a^b y^2\,dx$$

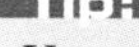

Tip: You must learn this formula.

There are several ways of describing the rotation, such as:

- rotated completely
- rotated through 360°
- rotated through 2π radians
- rotated through four right angles.

Example 5.9 The diagram shows the curve $y = \frac{1}{\sqrt{2x+1}}$, $x > -\frac{1}{2}$.

The shaded region R, bounded by the curve, the x-axis and the lines $x = 1$ and $x = 2$, is rotated completely about the x-axis. Find the exact value of the volume of the solid formed.

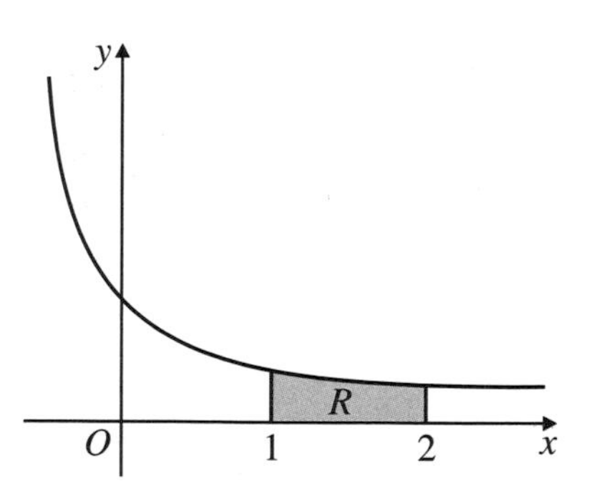

Step 1: Apply the volume formula.

$$V = \pi \int_a^b y^2\,dx$$

Step 2: Integrate with respect to x.

$$= \pi \int_1^2 \frac{1}{2x+1}\,dx$$

Tip: A common error is to forget to include π.

Step 3: Substitute the limits and evaluate.

$$= \pi \left[\tfrac{1}{2}\ln|2x+1|\right]_1^2$$

Tip: Recognise that the numerator is a multiple of the derivative of the denominator.

$$= \tfrac{1}{2}\pi(\ln|5| - \ln|3|)$$

$$= \tfrac{1}{2}\pi(\ln 5 - \ln 3)$$

$$= \tfrac{1}{2}\pi \ln \tfrac{5}{3}$$

Tip: You are asked for the exact value, so leave your answer in terms of π and $\ln a$.

Example 5.10 The curves $y = x^2$ and $y = \sqrt{x}$ are shown in the diagram. They intersect at (0, 0) and (1, 1). The region enclosed between the curves, shown shaded, is rotated through four right angles about the x-axis.

Note: This means that it is rotated completely, through 360°.

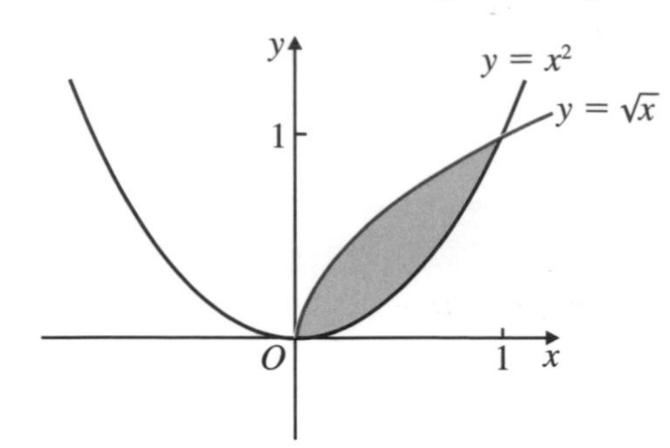

Show that the volume of the solid formed is $\frac{3}{10}\pi$.

Step 1: Find the volume when the area 'under' each curve is rotated about the x-axis.

Consider the area under $y = \sqrt{x}$:

$$V_1 = \pi\int_a^b y^2\,dx = \pi\int_0^1 x\,dx = \pi\left[\tfrac{1}{2}x^2\right]_0^1 = \tfrac{1}{2}\pi(1-0) = \tfrac{1}{2}\pi$$

Consider the area under $y = x^2$:

$$V_2 = \pi\int_a^b y^2\,dx = \pi\int_0^1 x^4\,dx = \pi\left[\tfrac{1}{5}x^5\right]_0^1 = \tfrac{1}{5}\pi(1-0) = \tfrac{1}{5}\pi$$

Tip: Alternatively, you could work out $\pi\int_a^b (y_1^2 - y_2^2)\,dx$, provided that $y_1 > y_2$ for $a \leqslant x \leqslant b$.

Step 2: Subtract the volumes.

The required volume is $V_1 - V_2 = \frac{1}{2}\pi - \frac{1}{5}\pi = \frac{3}{10}\pi$.

Volume of revolution for curves expressed in parametric form

Consider a curve, $y = f(x)$, given in terms of a parameter, t, where x and y are functions of t.

The volume of revolution formed when a region R is rotated completely about the x-axis is given by

$$\pi\int_{t_1}^{t_2} y^2\frac{dx}{dt}\,dt$$

Note: t_1 and t_2 are the limits of the parameter for the section of the curve being traced out.

Example 5.11 The diagram shows part of the graph of the curve given by the parametric equations $x = 3t^2$, $y = 2t + 1$, where $t \geqslant 0$.

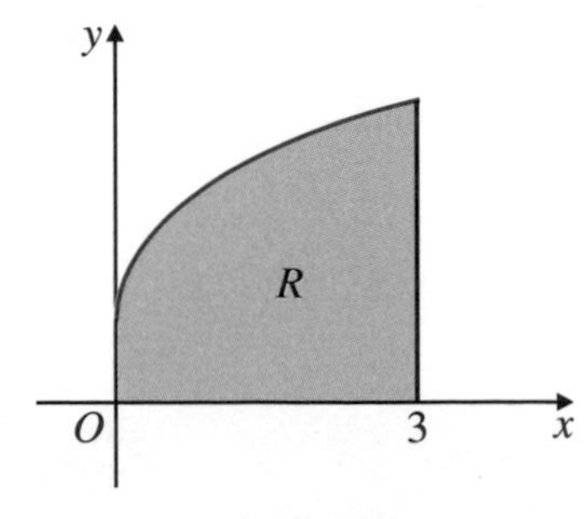

The region R enclosed by the curve, the coordinate axes and the line $x = 3$ is shown shaded in the diagram.

The volume of the solid generated when the region R is rotated completely about the x-axis is $k\pi$, where k is a positive integer. Find the value of k.

Step 1: State the volume formula in terms of the parameter t.

$$V = \pi\int_{t_1}^{t_2} y^2 \frac{dx}{dt}\,dt$$

Step 2: Work out the limits for t.

When $x = 0$, $3t^2 = 0 \Rightarrow t = 0$

When $x = 3$, $3t^2 = 3 \Rightarrow t = 1$

Tip: $t = -1$ is a solution of the equation $3t^2 = 3$, but is not applicable here, since $t \geqslant 0$.

Step 3: Work out y^2 and $\frac{dx}{dt}$.

Now $y = 2t + 1 \Rightarrow y^2 = (2t+1)^2 = 4t^2 + 4t + 1$

$$x = 3t^2 \Rightarrow \frac{dx}{dt} = 6t$$

Step 4: Substitute into the volume formula and integrate with respect to t.

So $$V = \pi\int_0^1 (4t^2 + 4t + 1) \times 6t\,dt$$

$$= 6\pi\int_0^1 (4t^3 + 4t^2 + t)\,dt$$

Tip: Where possible, take out numerical factors before integrating.

Step 5: Substitute the limits and evaluate.

$$= 6\pi\left[t^4 + \tfrac{4}{3}t^3 + \tfrac{1}{2}t^2\right]_0^1$$

$$= 6\pi\left(1 + \tfrac{4}{3} + \tfrac{1}{2} - 0\right)$$

$$= 17\pi$$

So $k = 17$.

SKILLS CHECK **5B: Volume of revolution**

1 The diagram shows a sketch of the curve $y = \dfrac{1}{x}$, $x > 0$.

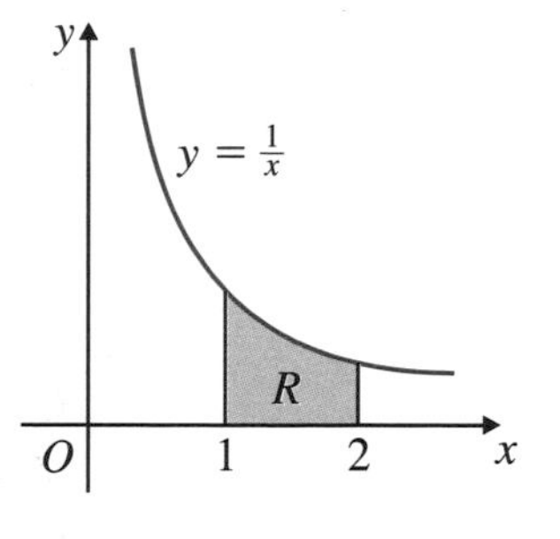

The region R is bounded by the x-axis and the lines $x = 1$ and $x = 2$.

a Find the exact value of the area of R.

b Find the volume of the solid formed when R is rotated completely about the x-axis.

2 The region bounded by the curve $y = e^x$, the coordinate axes and the line $x = 1$ is rotated completely about the x-axis. Find the exact value of the volume of the solid formed.

3 The region bounded by the curve $y = x^3$, the x-axis and the lines $x = 1$ and $x = 2$ is rotated through 360° about the x-axis. Find the exact value of the volume of the solid formed.

4 The diagram shows a sketch of the curve $y = \dfrac{1}{\sqrt{x+1}}$, $x > -1$.

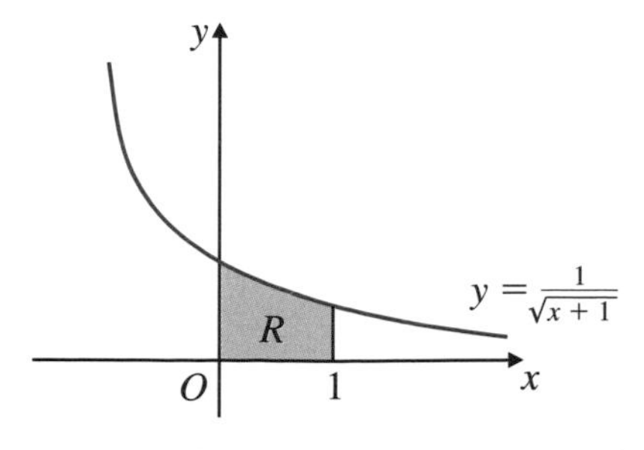

The region R, enclosed by the curve, the coordinate axes and the line $x = 1$, is rotated completely about the x-axis. Find the exact value of the volume of the solid formed.

5 The diagram shows the curve $y = x^2 + 1$ and the line $y = x + 1$. The line and the curve intersect at A and B.

a Find the coordinates of A and B.

The region enclosed between the curve and the line is rotated completely about the x-axis.

b Find the exact value of the volume of the solid formed.

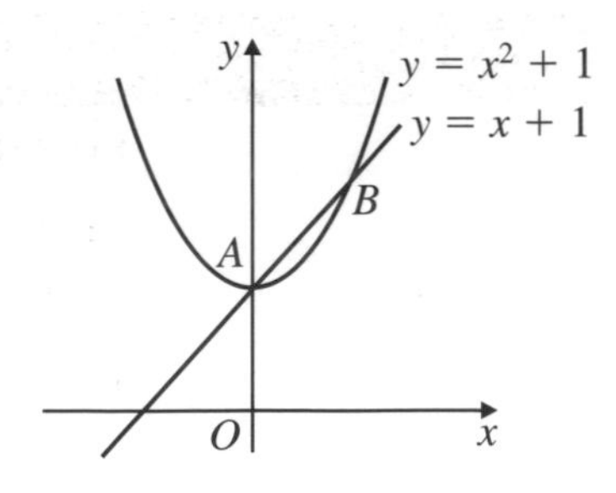

6 The diagram shows the curve $y = x(4 - x)$ and the line $y = x$.

The line and the curve intersect at the origin and at P.

a Find the coordinates of P.

The region enclosed between the curve and the line is rotated completely about the x-axis.

b Show that the volume of the solid formed is $\frac{108}{5}\pi$.

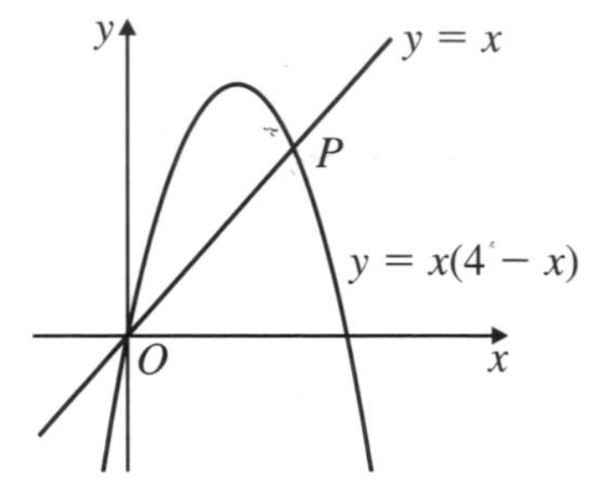

7 **a** Sketch the graph of $y = 9 - x^2$, indicating the coordinates of the points where the graph crosses the coordinate axes.

b The region between the curve and the x-axis from $x = -2$ to $x = -1$ is rotated through 360° about the x-axis. Find the volume of the solid generated, giving your answer to three significant figures.

8 The region bounded by the x-axis and the curve $y = \sin x$, for $0 \leqslant x \leqslant \pi$, is rotated completely about the x-axis. Find the exact value of the volume of the solid formed.

9 The diagram shows part of the curve given by the parametric equations $x = 3t$, $y = t^2 + 1$, where $t \geqslant 0$.

a Find the value of t when $x = 3$ and the value of t when $x = 6$.

The region enclosed by the curve, the x-axis and the lines $x = 3$ and $x = 6$ is rotated through 360° about the x-axis.

b Find the volume of the solid generated, giving your answer to three significant figures.

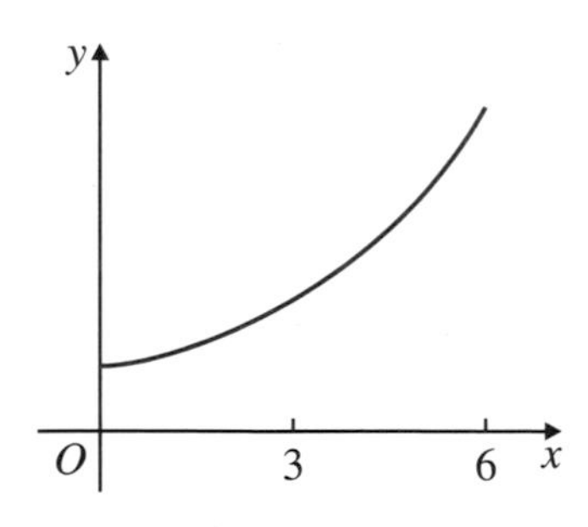

10 The diagram shows the curve given parametrically by the equations $x = \sin t$, $y = \sqrt{\cos t}$, $-\frac{1}{2}\pi \leqslant t \leqslant \frac{1}{2}\pi$. The region enclosed by the curve and the x-axis is rotated completely about the x-axis.

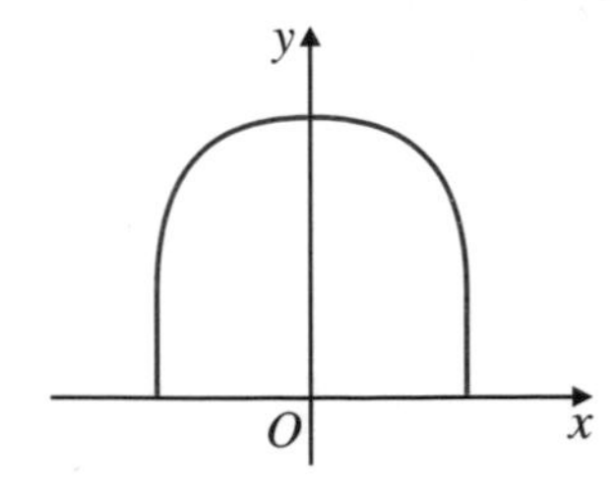

Find the exact value of the volume of the solid generated.

SKILLS CHECK **5B EXTRA is on the CD**

5.3 Integration by substitution

Simple cases of integration by substitution. This method as the reverse process of the chain rule.

Recall: Chain rule for differentiating a function of a function (C3 Section 4.1).

Many of the integration results stated earlier have been derived using the reverse process of the chain rule, known as **integration by substitution**. Using the substitution u, where u is a function of x,

$$y = \int f(x)\,dx$$

becomes

$$y = \int f(x)\frac{dx}{du}\,du$$

Example 5.12 Use the substitution $u = 2x - 1$ to find $\int (2x - 1)^8\,dx$.

Step 1: Use the substitution to find $\frac{dx}{du}$.

Let $u = 2x - 1$, then $\frac{du}{dx} = 2 \Rightarrow \frac{dx}{du} = \frac{1}{2}$

Recall: $\frac{dx}{du} = \frac{1}{\frac{du}{dx}}$ (C3 Section 4.3).

Step 2: Rewrite the integral so that it is with respect to u.

$$\int (2x - 1)^8\,dx = \int (2x - 1)^8 \frac{dx}{du}\,du$$

Step 3: Substitute to get an integral in terms of u.

$$= \int u^8 \times \tfrac{1}{2}\,du$$

Step 4: Integrate with respect to u.

$$= \tfrac{1}{2}\int u^8\,du$$

$$= \tfrac{1}{2} \times \tfrac{1}{9}u^9 + c$$

Step 5: Rewrite in terms of x.

$$= \tfrac{1}{18}(2x - 1)^9 + c$$

Tip: You could find $\frac{dx}{du}$ by making x the subject, where $x = \frac{u+1}{2} = \frac{1}{2}u + \frac{1}{2}$

Example 5.13 Use the substitution $u = 3 - 2x$ to find $\int e^{3-2x}\,dx$.

Step 1: Use the substitution to find $\frac{dx}{du}$.

Let $u = 3 - 2x$, then $\frac{du}{dx} = -2 \Rightarrow \frac{dx}{du} = -\frac{1}{2}$

Step 2: Rewrite the integral so that it is with respect to u.

$$\int e^{3-2x}\,dx = \int e^{3-2x}\frac{dx}{du}\,du$$

Step 3: Substitute to get an integral in terms of u.

$$= \int e^u \times (-\tfrac{1}{2})\,du$$

Step 4: Integrate with respect to u.

$$= -\tfrac{1}{2}\int e^u\,du$$

$$= -\tfrac{1}{2}e^u + c$$

Step 5: Rewrite in terms of x.

$$= -\tfrac{1}{2}e^{3-2x} + c$$

Example 5.14 Use the substitution $u = x^3 + 8$ to find $\int x^2(x^3 + 8)^5\,dx$.

Step 1: Use the substitution to find $\frac{dx}{du}$.

Let $u = x^3 + 8$, then $\frac{du}{dx} = 3x^2 \Rightarrow \frac{dx}{du} = \frac{1}{3x^2} \Rightarrow x^2\frac{dx}{du} = \frac{1}{3}$

Step 2: Rewrite the integral so that it is with respect to u.

$$\int x^2(x^3 + 8)^5\,dx = \int x^2(x^3 + 8)^5\frac{dx}{du}\,du$$

Step 3: Substitute to get an integral in terms of u.

$$= \int (x^3 + 8)^5 x^2 \frac{dx}{du}\,du$$

$$= \int u^5 \times \tfrac{1}{3}\,du$$

Tip: Substitute for $x^2\frac{dx}{du}$ in the integral.

Step 4: Integrate with respect to u.

$$= \tfrac{1}{3}\int u^5\,du$$

$$= \tfrac{1}{3} \times \tfrac{1}{6}u^6 + c$$

Step 5: Rewrite in terms of x.

$$= \tfrac{1}{18}(x^3 + 8)^6 + c$$

Example 5.15 Use the substitution $u = 2x + 1$ to find $\int x\sqrt{2x+1}\,dx$.

Step 1: Use the substitution to find x and $\frac{dx}{du}$.

Let $u = 2x + 1$, then $\frac{du}{dx} = 2 \Rightarrow \frac{dx}{du} = \frac{1}{2}$

Also $x = \frac{u-1}{2} = \frac{1}{2}(u-1)$

Note: You need to substitute for x, so write x in terms of u.

Step 2: Rewrite the integral so that it is with respect to u.

$$\int x\sqrt{2x+1}\,dx = \int x\sqrt{2x+1}\,\frac{dx}{du}\,du$$

Step 3: Substitute to get an integral in terms of u.

$$= \int \tfrac{1}{2}(u-1)u^{\frac{1}{2}} \times \tfrac{1}{2}\,du$$

$$= \tfrac{1}{4}\int (u^{\frac{3}{2}} - u^{\frac{1}{2}})\,du$$

Step 4: Integrate with respect to u.

$$= \tfrac{1}{4} \times \left(\frac{1}{\frac{5}{2}}u^{\frac{5}{2}} - \frac{1}{\frac{3}{2}}u^{\frac{3}{2}}\right) + c$$

Step 5: Rewrite in terms of x.

$$= \tfrac{1}{4} \times (\tfrac{2}{5}(2x+1)^{\frac{5}{2}} - \tfrac{2}{3}(2x+1)^{\frac{3}{2}}) + c$$

$$= \tfrac{1}{10}(2x+1)^{\frac{5}{2}} - \tfrac{1}{6}(2x+1)^{\frac{3}{2}} + c$$

Example 5.16 Use the substitution $x = 2\tan u$ to find $\int \frac{1}{4+x^2}\,dx$.

Step 1: Use the substitution to find $\frac{dx}{du}$.

Let $x = 2\tan u$, then $\frac{dx}{du} = 2\sec^2 u$

Note: x is a function of u, so you will get $\frac{dx}{du}$ straight away when you differentiate.

Step 2: Simplify the expression in x.

Also $4 + x^2 = 4 + 4\tan^2 u = 4(1 + \tan^2 u) = 4\sec^2 u$

Recall: $1 + \tan^2 A \equiv \sec^2 A$

Step 3: Rewrite the integral so that it is with respect to u.

$$\int \frac{1}{4+x^2}\,dx = \int \frac{1}{4+x^2}\,\frac{dx}{du}\,du$$

Step 4: Substitute to get an integral in terms of u.

$$= \int \frac{1}{4\sec^2 u} \times 2\sec^2 u\,du$$

$$= \int \tfrac{1}{2}\,du$$

Step 5: Integrate with respect to u.

$$= \tfrac{1}{2}u + c$$

Step 6: Rewrite in terms of x.

$x = 2\tan u \Rightarrow \tan u = \frac{x}{2} \Rightarrow u = \arctan\left(\frac{x}{2}\right)$

Recall: Inverse trigonometric functions (C3 Section 2.1).

so $\int \frac{1}{4+x^2}\,dx = \tfrac{1}{2}\arctan\left(\frac{x}{2}\right) + c$

When performing **definite integration** with respect to x, you can do either of the following:

- work through the integration with no limits, obtaining an answer in terms of x, and then substitute the x-limits,
- change the limits to the new variable and evaluate, using these new limits.

These methods are illustrated in the next two examples.

Example 5.17 **a** Use the substitution $u = \sin x$ to find $\int \sin^3 x \cos x \, dx$.

b Hence evaluate $\int_0^{\frac{1}{2}\pi} \sin^3 x \cos x \, dx$.

Step 1: Use the substitution to find x and $\frac{dx}{du}$.

a Let $u = \sin x$, then $\frac{du}{dx} = \cos x \Rightarrow \cos x \frac{dx}{du} = 1$

Step 2: Rewrite the integral so that it is with respect to u.

$$\int \sin^3 x \cos x \, dx = \int \sin^3 x \cos x \frac{dx}{du} \, du$$

Step 3: Substitute to get an integral in terms of u.

$$= \int u^3 \times 1 \, du$$

Step 4: Integrate with respect to u.

$$= \tfrac{1}{4} u^4 + c$$

Step 5: Rewrite in terms of x.

$$= \tfrac{1}{4} \sin^4 x + c$$

Step 1: Using your answer to part **a**, substitute the limits and evaluate.

b
$$\int_0^{\frac{1}{2}\pi} \sin^3 x \cos x \, dx = \left[\tfrac{1}{4} \sin^4 x\right]_0^{\frac{1}{2}\pi}$$
$$= \tfrac{1}{4} \sin^4 (\tfrac{1}{2}\pi) - \tfrac{1}{4} \sin^4 0$$
$$= \tfrac{1}{4}$$

Tip: Exam questions often split the question into two parts like this.

Tip: Substitute for $\cos x \frac{dx}{du}$ in the integral.

Note: When substituting the limits, the integration constant c is omitted.

Example 5.18 Use the substitution $u = 3x + 1$ to show that

$$\int_1^2 \frac{x}{3x+1} \, dx = p + q \ln r$$

where p, q and r are positive rational numbers to be found.

Step 1: Use the substitution to find x and $\frac{dx}{du}$.

Let $u = 3x + 1$, then $\frac{du}{dx} = 3 \Rightarrow \frac{dx}{du} = \frac{1}{3}$

Also $x = \frac{u-1}{3} = \frac{1}{3}(u - 1)$

Step 2: Work out the limits for u.

Limits:

x	1	2
u	4	7

Step 3: Rewrite the integral so that it is with respect to u.

$$\int_1^2 \frac{x}{3x+1} \, dx = \int_{x=1}^{x=2} \frac{x}{3x+1} \frac{dx}{du} \, du$$

Step 4: Substitute to get an integral in terms of u, with u-limits.

$$= \int_{u=4}^{u=7} \frac{\frac{1}{3}(u-1)}{u} \times \tfrac{1}{3} \, du$$
$$= \tfrac{1}{9} \int_4^7 \left(1 - \frac{1}{u}\right) du$$

Step 5: Integrate with respect to u and evaluate.

$$= \tfrac{1}{9} \Big[u - \ln|u|\Big]_4^7$$
$$= \tfrac{1}{9}(7 - \ln 7 - (4 - \ln 4))$$
$$= \tfrac{1}{9}(7 - \ln 7 - 4 + \ln 4)$$
$$= \tfrac{1}{9}(3 + \ln \tfrac{4}{7})$$

Step 6: Express the answer in the required format and state the values of p, q and r.

$$= \tfrac{1}{3} + \tfrac{1}{9} \ln \tfrac{4}{7}$$

Hence $p = \frac{1}{3}$, $q = \frac{1}{9}$ and $r = \frac{4}{7}$.

Tip: When $x = 1$, $u = 3 \times 1 + 1 = 4$. When $x = 2$, $u = 3 \times 2 + 1 = 7$.

Tip: Take out any numerical factors before integrating and/or before substituting the limits.

SKILLS CHECK **5C: Integration by substitution**

1 Use the substitution $u = 3x + 1$ to find the exact value of

a $\int_0^1 (3x+1)^5\,dx$ **b** $\int_1^5 \sqrt{3x+1}\,dx$

2 By using the substitution $u = x - 4$, or otherwise,

a show that $\int_5^8 \frac{1}{x-4}\,dx = 2\ln 2$ **b** evaluate $\int_5^8 \frac{1}{(x-4)^2}\,dx$

3 Using the substitution $u = 2x + 1$, show each of the following:

a $\int x(2x+1)^3\,dx = \frac{1}{80}(8x-1)(2x+1)^4 + c$ **b** $\int_0^1 \frac{x}{(2x+1)^3}\,dx = \frac{1}{18}$

4 By using the substitution $u = x^2 + 2$, or otherwise, find $\int xe^{x^2+2}\,dx$.

5 Use the substitution $u = 9 - x$ to find $\int \frac{3}{\sqrt{9-x}}\,dx$.

6 Using the substitution $u = e^{2x}$, or otherwise, find $\int \frac{e^{2x}}{1-e^{2x}}\,dx$.

7 Use the substitution $u = 1 + \ln x$ to find $\int \frac{1+\ln x}{x}\,dx$.

8 **a** Using the substitution $u = 2x + 1$, find $\int \frac{x}{(2x+1)^2}\,dx$.

b Hence evaluate $\int_2^4 \frac{x}{(2x+1)^2}\,dx$, giving your answer to three significant figures.

9 **a** Use the substitution $u = \cos x$ to find $\int \cos^2 x \sin x\,dx$.

b Hence evaluate $\int_0^{\frac{1}{2}\pi} \cos^2 x \sin x\,dx$.

10 Use the substitution $u = 1 + \cos x$ to find $\int_0^{\pi} \frac{\sin^3 x}{1+\cos x}\,dx$.

11 Using the substitution $x = \sin u$, find $\int \frac{1}{\sqrt{1-x^2}}\,dx$.

SKILLS CHECK **5C EXTRA is on the CD**

5.4 Integration by parts

Simple cases of integration by parts. This method as the reverse process of the product rule.

The method known as **integrating by parts** is useful for integrating a **product** of two functions of x. One function is denoted by u and the other function is denoted by $\frac{dv}{dx}$, where

$$\int u\frac{dv}{dx}\,dx = uv - \int v\frac{du}{dx}\,dx$$

Note:
This formula has been obtained from the product rule for differentiating (C3 Section 4.2).

Example 5.19 Use integration by parts to find:

a $\int xe^{3x}\,dx$ **b** $\int x^2e^{3x}\,dx$

Step 1: Decide on the functions u and $\frac{dv}{dx}$, and find $\frac{du}{dx}$ and v.

a Let $u = x$, then $\frac{du}{dx} = 1$

Let $\frac{dv}{dx} = e^{3x}$, then $v = \int e^{3x}\,dx = \frac{1}{3}e^{3x}$

Step 2: Apply the formula for integrating by parts.

$$\int u\frac{dv}{dx}\,dx = uv - \int v\frac{du}{dx}\,dx$$

so $\int xe^{3x}\,dx = x \times \frac{1}{3}e^{3x} - \int \frac{1}{3}e^{3x} \times 1\,dx$

$$= \tfrac{1}{3}xe^{3x} - \tfrac{1}{3} \times \tfrac{1}{3}e^{3x} + c$$
$$= \tfrac{1}{3}xe^{3x} - \tfrac{1}{9}e^{3x} + c$$

Note:
$\int e^{kx}\,dx = \frac{1}{k}e^{kx} + c$, but the integration constant c is omitted at this stage.

Note:
Include the integration constant at the final stage when finding an indefinite integral.

Step 1: Decide on the functions u and $\frac{dv}{dx}$, and find $\frac{du}{dx}$ and v.

b Let $u = x^2$, then $\frac{du}{dx} = 2x$

Let $\frac{dv}{dx} = e^{3x}$, then $v = \int e^{3x}\,dx = \frac{1}{3}e^{3x}$

Step 2: Apply the formula for integrating by parts.

$$\int x^2e^{3x}\,dx = x^2 \times \tfrac{1}{3}e^{3x} - \int \tfrac{1}{3}e^{3x} \times 2x\,dx$$

Step 3: Apply the formula again.

$$= \tfrac{1}{3}x^2e^{3x} - \tfrac{2}{3}\int xe^{3x}\,dx$$
$$= \tfrac{1}{3}x^2e^{3x} - \tfrac{2}{3}\left(\tfrac{1}{3}xe^{3x} - \tfrac{1}{9}e^{3x}\right) + c$$
$$= \tfrac{1}{3}x^2e^{3x} - \tfrac{2}{9}xe^{3x} + \tfrac{2}{27}e^{3x} + c$$

Note:
You have to apply integration by parts again.

Tip:
Notice the connection with part **a**. You can quote your result. Just put one integration constant at the end.

Example 5.20 **a** Find $\int (3x + 1)\sin 2x\,dx$.

b Hence evaluate $\int_0^{\frac{1}{2}\pi} (3x + 1)\sin 2x\,dx$.

Note:
This is a common approach in a question, where you are asked to find an indefinite integral first, then to evaluate the definite integral.

Step 1: Decide on the functions u and $\frac{dv}{dx}$, and find $\frac{du}{dx}$ and v.

a Let $u = 3x + 1$, then $\frac{du}{dx} = 3$

Let $\frac{dv}{dx} = \sin 2x$, then $v = \int \sin 2x\,dx = -\frac{1}{2}\cos 2x$

Tip:
$\int \sin kx\,dx = -\frac{1}{k}\cos kx + c$, but leave out the integration constant until the final stage.

Step 2: Apply the formula for integrating by parts and tidy up the expressions.

$$\int (3x + 1)\sin 2x\,dx = (3x + 1) \times \left(-\tfrac{1}{2}\cos 2x\right) - \int \left(-\tfrac{1}{2}\cos 2x\right) \times 3\,dx$$
$$= -\tfrac{1}{2}(3x + 1)\cos 2x + \tfrac{3}{2}\int \cos 2x\,dx$$
$$= -\tfrac{1}{2}(3x + 1)\cos 2x + \tfrac{3}{2} \times \tfrac{1}{2}\sin 2x + c$$
$$= -\tfrac{1}{2}(3x + 1)\cos 2x + \tfrac{3}{4}\sin 2x + c$$

Step 3: Apply the limits to the answer for the indefinite integral and evaluate.

b $\int_0^{\frac{1}{2}\pi} (3x + 1)\sin 2x\,dx$

$$= \left[-\tfrac{1}{2}(3x + 1)\cos 2x + \tfrac{3}{4}\sin 2x\right]_0^{\frac{1}{2}\pi}$$
$$= -\tfrac{1}{2}\left(\tfrac{3}{2}\pi + 1\right)\cos \pi + \tfrac{3}{4}\sin \pi - \left(-\tfrac{1}{2}(3 \times 0 + 1)\cos 0 + \tfrac{3}{4}\sin 0\right)$$
$$= -\tfrac{1}{2}\left(\tfrac{3}{2}\pi + 1\right)(-1) + 0 - \left(-\tfrac{1}{2} \times 1 + 0\right)$$
$$= \tfrac{3}{4}\pi + \tfrac{1}{2} + \tfrac{1}{2}$$
$$= \tfrac{3}{4}\pi + 1$$

Tip:
To evaluate this **definite** integral, use your answer to part **a**, omit c and substitute the limits.

When evaluating a **definite integral**, if you are not asked to find the indefinite integral first, you can substitute the limits as you go along, using the result

$$\int_a^b u\frac{dv}{dx}\,dx = \Big[uv\Big]_a^b - \int_a^b v\frac{du}{dx}\,dx$$

This is illustrated in the following example.

Tip: If you use this method, you will need to take care; do not try to do too many things at once.

Example 5.21 Evaluate $\int_0^{\frac{1}{2}\pi} x\cos x\,dx$.

Step 1: Decide on the functions u and $\frac{dv}{dx}$, and find $\frac{du}{dx}$ and v.

Let $u = x$, then $\frac{du}{dx} = 1$

Let $\frac{dv}{dx} = \cos x$, then $v = \int \cos x\,dx = \sin x$

Step 2: Apply the formula for integrating by parts.

$$\int_0^{\frac{1}{2}\pi} x\cos x\,dx = \Big[x\sin x\Big]_0^{\frac{1}{2}\pi} - \int_0^{\frac{1}{2}\pi} \sin x \times 1\,dx$$

Step 3: Substitute the limits for the first part.

$$= (\tfrac{1}{2}\pi\sin\tfrac{1}{2}\pi - 0) - \int_0^{\frac{1}{2}\pi}\sin x\,dx$$

Step 4: Integrate the second part.

$$= \tfrac{1}{2}\pi - \Big[-\cos x\Big]_0^{\frac{1}{2}\pi}$$

$$= \tfrac{1}{2}\pi + \Big[\cos x\Big]_0^{\frac{1}{2}\pi}$$

Step 5: Substitute the limits for the second part.

$$= \tfrac{1}{2}\pi + (\cos\tfrac{1}{2}\pi - \cos 0)$$

$$= \tfrac{1}{2}\pi - 1$$

Tip: $\cos 0 = 1$. Do not forget to consider this.

Choice of u and $\frac{dv}{dx}$

If the product involves a polynomial in x, then this is usually taken to be u. However, the exception to this is when the other function involves $\ln x$. This is illustrated in the following example.

Tip: If the polynomial involves x^2, then integration by parts needs to be carried out twice.

Example 5.22 Find **a** $\int x\ln x\,dx$ **b** $\int \ln x\,dx$

Step 1: Write the integral with $\ln x$ first.

a $\int x\ln x\,dx = \int(\ln x)\times x\,dx$

Step 2: Decide on the functions u and $\frac{dv}{dx}$, and find $\frac{du}{dx}$ and v.

Let $u = \ln x$, then $\frac{du}{dx} = \frac{1}{x}$

Let $\frac{dv}{dx} = x$, then $v = \int x\,dx = \frac{1}{2}x^2$

Step 3: Apply the formula for integrating by parts and tidy up the expressions.

$$\int(\ln x)\times x\,dx = \ln x\times\tfrac{1}{2}x^2 - \int\tfrac{1}{2}x^2\times\frac{1}{x}\,dx$$

$$= \tfrac{1}{2}x^2\ln x - \tfrac{1}{2}\int x\,dx$$

$$= \tfrac{1}{2}x^2\ln x - \tfrac{1}{2}\times\tfrac{1}{2}x^2 + c$$

$$= \tfrac{1}{2}x^2\ln x - \tfrac{1}{4}x^2 + c$$

Tip: You now have to integrate a simple function of x.

Step 1: Write the integral as a product with $\ln x$ first.

b $\int \ln x \, dx = \int (\ln x) \times 1 \, dx$

Step 2: Decide on the functions u and $\frac{dv}{dx}$, and find $\frac{du}{dx}$ and v.

Let $u = \ln x$, then $\frac{du}{dx} = \frac{1}{x}$

Let $\frac{dv}{dx} = 1$, then $v = \int 1 \, dx = x$

Step 3: Apply the formula for integrating by parts and tidy up the expressions.

$$\int (\ln x) \times 1 \, dx = \ln x \times x - \int x \times \frac{1}{x} \, dx$$

$$= x \ln x - \int 1 \, dx$$

$$= x \ln x - x + c$$

Hence $\int \ln x \, dx = x \ln x - x + c$.

Tip: Set up the product by multiplying by 1.

SKILLS CHECK 5D: Integration by parts

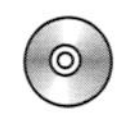

1 Use integration by parts to find

a $\int (2x + 3) \cos x \, dx$ **b** $\int x \cos (2x + 3) \, dx$

2 Use integration by parts to find

a $\int x e^{3x} \, dx$ **b** $\int e^{3x} (x - 1) \, dx$

3 Evaluate $\int_2^3 x^2 \ln x \, dx$, giving your answer to three significant figures.

4 **a** Find $\int 5x \sin 2x \, dx$ **b** Hence show that $\int_0^{\frac{1}{2}\pi} x(1 + 5 \sin 2x) \, dx = \frac{1}{8}\pi^2 + \frac{5}{4}\pi$

5 **a** **i** Find $\int_0^{\frac{1}{2}\pi} x \sin x \, dx$ **ii** Find $\int_0^{\frac{1}{2}\pi} x^2 \cos x \, dx$

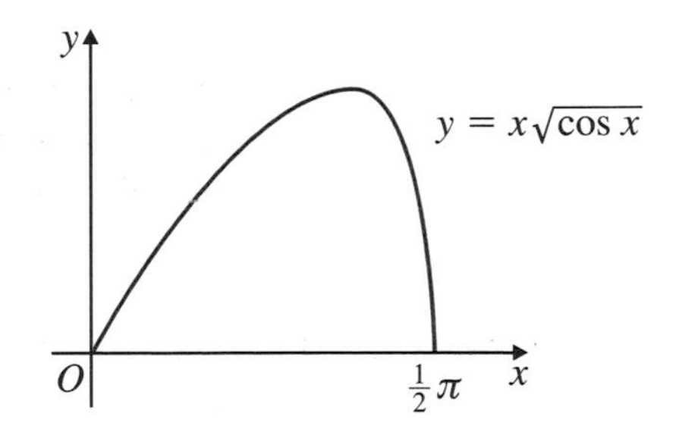

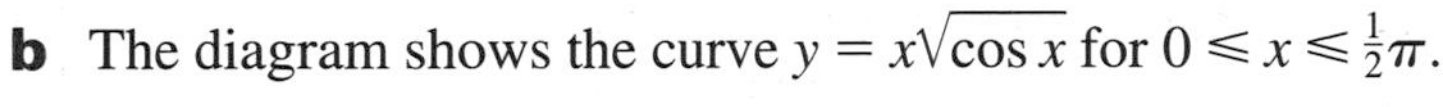

b The diagram shows the curve $y = x\sqrt{\cos x}$ for $0 \leqslant x \leqslant \frac{1}{2}\pi$.

The region bounded by the curve and the x-axis is rotated through 360° about the x-axis.

Find the exact value of the volume of the solid generated.

6 **a** Express $\sin^2 A$ in terms of $\cos 2A$.

b Evaluate $\int_0^{\pi} x \sin^2 x \, dx$.

7 **a** Use integration by parts to find $\int x e^{-2x} \, dx$.

b The diagram shows the curve $y = xe^{-x}$.

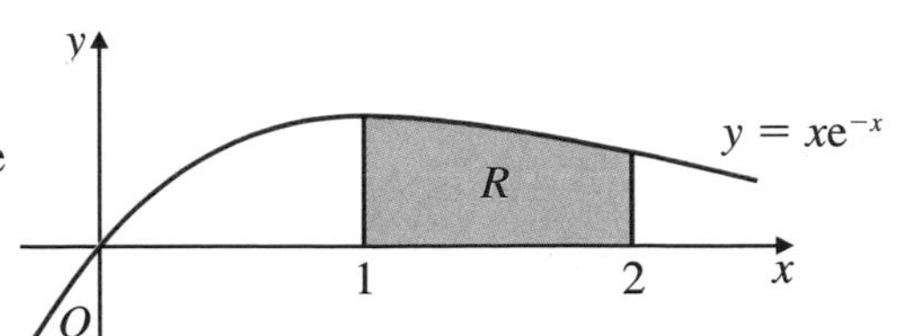

The region R, bounded by the curve, the x-axis and the lines $x = 1$ and $x = 2$, is rotated completely about the x-axis.

Show that the volume of the solid generated is $\frac{\pi(5e^2 - 13)}{4e^4}$.

8 **a** Use integration by parts to find $\int x \sec^2 x \, dx$.

The diagram shows part of the curve of $y = \sqrt{x} \sec x$ and the line $x = \frac{1}{4}\pi$. The region enclosed by the curve, the line and the x-axis is rotated completely about the x-axis.

b Show that the volume of the solid generated is $\frac{1}{4}\pi^2 - \frac{1}{2}\pi \ln 2$.

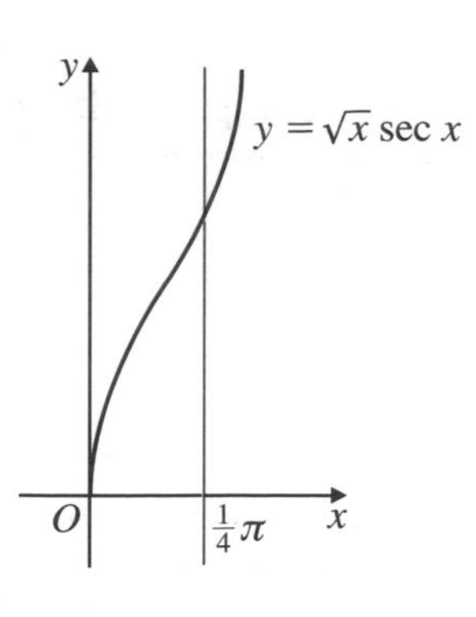

SKILLS CHECK **5D EXTRA is on the CD**

5.5 Integration using partial fractions

Simple cases of integration using partial fractions.

When you are asked to integrate a rational function where the denominator factorises, a useful method to investigate is whether it can be split into partial fractions. If so, the function may be easier to integrate in this partial fraction form.

Recall: Partial fractions (Section 1.1).

In Example 1.1 it was shown that

$$\frac{2x-4}{(x-3)(x+1)} \equiv \frac{1}{2(x-3)} + \frac{3}{2(x+1)}$$

Tip: These two partial fractions can now be integrated separately.

So $\displaystyle\int \frac{2x-4}{(x-3)(x+1)} \, dx = \int \left(\frac{1}{2(x-3)} + \frac{3}{2(x+1)} \right) dx$

$$= \int \left(\frac{1}{2}\left(\frac{1}{x-3}\right) + \frac{3}{2}\left(\frac{1}{x+1}\right) \right) dx$$

$$= \tfrac{1}{2} \ln|x-3| + \tfrac{3}{2} \ln|x+1| + c$$

Tip: Spot the use of $\int \frac{f'(x)}{f(x)} \, dx = \ln|f(x)| + c$ (Section 5.1).

Example 5.23 **a** Express $\dfrac{1}{x^2(x+1)}$ in partial fractions.

b Hence show that $\displaystyle\int_1^2 \frac{1}{x^2(x+1)} \, dx = \frac{1}{2} + \ln\frac{3}{4}$.

Step 1: Set out the partial fractions.

a Let $\dfrac{1}{x^2(x+1)} \equiv \dfrac{A}{x} + \dfrac{B}{x^2} + \dfrac{C}{x+1}$

Recall: Format when there are repeated factors in the denominator (Section 1.1).

Step 2: Add the fractions.

$$\frac{1}{x^2(x+1)} \equiv \frac{Ax(x+1) + B(x+1) + Cx^2}{x^2(x+1)}$$

Step 3: Equate the numerators.

So $\quad 1 \equiv Ax(x+1) + B(x+1) + Cx^2$

Tip: Don't expand the brackets here.

Step 4: Substitute appropriate values or compare coefficients.

Substituting $x = 0$,

$$1 = A \times 0 + B \times 1 + C \times 0$$

$\Rightarrow \quad B = 1$

Substituting $x = -1$,

$$1 = A \times 0 + B \times 0 + C \times (-1)^2$$

$\Rightarrow \quad C = 1$

Tip: Substitute $x = -1$ because then the factor $(x + 1)$ is equal to zero.

Equating coefficients of x^2,

$$0 = A + C$$

$\Rightarrow \quad A = -1$

Step 5: Write out the partial fractions.

$$\frac{1}{x^2(x+1)} \equiv -\frac{1}{x} + \frac{1}{x^2} + \frac{1}{x+1}$$

Step 1: Integrate the partial fractions separately.

b $\displaystyle\int_1^2 \frac{1}{x^2(x+1)}\,dx = \int_1^2 \left(-\frac{1}{x} + \frac{1}{x^2} + \frac{1}{x+1}\right) dx$

$\displaystyle = \int_1^2 \left(-\frac{1}{x} + x^{-2} + \frac{1}{x+1}\right) dx$

Tip: Write x^n in index form (except when $n = -1$).

Step 2: Substitute the limits.

$\displaystyle = \Big[-\ln|x| - x^{-1} + \ln|x+1|\Big]_1^2$

Tip: Recognise the log integrals (Section 5.1).

Step 3: Arrange into the required format.

$= -\ln 2 - 2^{-1} + \ln 3 - (-\ln 1 - 1^{-1} + \ln 2)$

$= -\ln 2 - \frac{1}{2} + \ln 3 + 1 - \ln 2$

$= \frac{1}{2} + \ln 3 - 2\ln 2$

$= \frac{1}{2} + \ln 3 - \ln 4$

$= \frac{1}{2} + \ln\frac{3}{4}$

Recall: Log laws
$n \log a = \log a^n$
$\log a - \log b = \log\frac{a}{b}$
(C2 Section 5.2).

SKILLS CHECK **5E: Integration using partial fractions**

1 $\displaystyle f(x) \equiv \frac{x-11}{(3x+1)(2x-5)} \equiv \frac{A}{3x+1} + \frac{B}{2x-5}.$

a Find the values of the constants A and B.

b Hence find $\displaystyle\int \frac{x-11}{(3x+1)(2x-5)}\,dx.$

c Hence show that $\displaystyle\int_1^2 \frac{x-11}{(3x+1)(2x-5)}\,dx = \frac{2}{3}\ln\frac{7}{4} + \frac{1}{2}\ln 3.$

2 **a** Express $\displaystyle\frac{2x^2-9x-31}{(x+2)(2x-1)(x+3)}$ in partial fractions.

b Hence evaluate $\displaystyle\int_{-1}^0 \frac{2x^2-9x-31}{(x+2)(2x-1)(x+3)}\,dx$, expressing your answer as a single natural logarithm.

3 **a** Given that $\displaystyle\frac{x^2}{x^2-9} \equiv A + \frac{B}{x+3} + \frac{C}{x-3}$, find the values of the constants A, B and C.

b Hence find $\displaystyle\int \frac{x^2}{x^2-9}\,dx.$

4 **a** Express $\displaystyle\frac{1}{x(x-1)^2}$ in partial fractions.

b Hence find $\displaystyle\int_2^3 \frac{1}{x(x-1)^2}\,dx.$

5 $\displaystyle f(x) \equiv \frac{16}{x^2(4-x)} = \frac{A}{x} + \frac{B}{x^2} + \frac{C}{4-x}.$

a Find the values of the constants A, B and C.

The diagram shows the graph $\displaystyle y = \frac{4}{x\sqrt{4-x}}$, $x > 0$.

The region R, enclosed by the curve, the x-axis and the lines $x = 1$ and $x = 3$, is rotated through 360° about the x-axis.

b Find the volume of the solid generated, giving your answer to three significant figures.

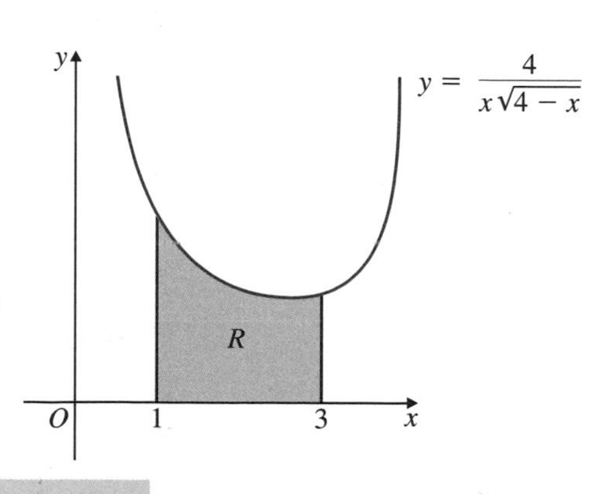

SKILLS CHECK **5E EXTRA is on the CD**

5.6 Differential equations

Analytical solution of simple first order differential equations with separable variables.

You solved the following type of first order differential equation in *Core 1*:

$$\text{If } \frac{dy}{dx} = f(x), \text{ then } y = \int f(x)\, dx$$

For example:

$$\text{If } \frac{dy}{dx} = 2x + 1, \text{ then } y = \int (2x + 1)\, dx = x^2 + x + c$$

This solution contains an integration constant (c) and is called the **general solution** of the differential equation.

If, however, we know that $y = 10$ when $x = 2$, we can find a **particular solution** as follows:

$$y = 10 \text{ when } x = 2 \Rightarrow 10 = 2^2 + 2 + c$$
$$c = 4$$

Hence the particular solution is $y = x^2 + x + 4$.

In *Core 4* the differential equations that you have to solve include ones such as

$$\frac{dy}{dx} = 3x^2y \quad \text{or} \quad e^x\frac{dy}{dx} = \frac{1}{y}$$

The expressions in x and y will be able to be separated into the form

$$f(y)\frac{dy}{dx} = g(x)$$

To solve, integrate both sides with respect to x to give

$$\int f(y)\frac{dy}{dx}\, dx = \int g(x)\, dx$$

This reduces to the form

$$\int f(y)\, dy = \int g(x)\, dx$$

Note:
A first order differential equation in x and y is an equation containing the differential coefficient $\frac{dy}{dx}$ but no higher order differential coefficients such as $\frac{d^2y}{dx^2}$.

Note:
A particular solution does not have an integration constant.

Note:
This is known as 'variables separable' form.

Note:
In practice, take all the terms in y to the side with dy and take all the terms to the other side with dx. This is known as **separating the variables**.

Example 5.24 Given that $\frac{dy}{dx} = \frac{x}{y^2}$, express y in terms of x.

Step 1: Separate the variables to get the form $f(y)\frac{dy}{dx} = g(x)$.

$$\frac{dy}{dx} = \frac{x}{y^2}$$

$$\Rightarrow \int y^2\, dy = \int x\, dx$$

Step 2: Integrate each side separately.

$$\Rightarrow \frac{1}{3}y^3 = \frac{1}{2}x^2 + c$$

Step 3: Make y the subject.

$$y^3 = \frac{3}{2}x^2 + k$$

$$y = \sqrt[3]{\frac{3}{2}x^2 + k}$$

Tip:
You can use any letter for the constant. Here $k = 3c$.

Example 5.25 It is given that $\frac{dy}{dx} = 3x^2y$.

a Show that $y = Ae^{x^3}$, where A is a constant.

b Given also that $y = 2$ when $x = 0$, find the value of A.

Step 1: Separate the variables to get the form $f(y)\frac{dy}{dx} = g(x)$.

a
$$\frac{dy}{dx} = 3x^2y$$
$$\Rightarrow \int \frac{1}{y}\,dy = \int 3x^2\,dx$$

Step 2: Integrate each side separately.

$$\Rightarrow \ln y = x^3 + c$$
$$y = e^{x^3+c}$$
$$= e^{x^3} \times e^c$$

Step 3: Make y the subject and write in the required form.

$$= Ae^{x^3}$$

Alternatively:

$$\ln y = x^3 + c$$
$$\ln y = x^3 + \ln A$$
$$\ln y - \ln A = x^3$$
$$\ln \frac{y}{A} = x^3$$
$$y = Ae^{x^3}$$

Step 4: Use the given condition to find the value of the constant.

b $y = 2$ when $x = 0$

$$\Rightarrow 2 = Ae^0$$
$$2 = A \times 1$$

Hence $A = 2$.

Note: It is better to leave numerical factors where they will be in the numerator of a fraction.

Recall: $\ln a = b \Leftrightarrow e^b = a$

Tip: Let $e^c = A$.

Tip: Let $c = \ln A$.

Recall: $\log a - \log b = \log\frac{a}{b}$

Recall: $e^0 = 1$

The differential equation may be set in the context of a problem, as in the following example.

Example 5.26 A mathematical model for the number of bacteria, N, in an experiment states that N is increasing at a rate proportional to the number of bacteria present at time t, where t is measured in minutes. Initially there are 1000 bacteria and after 5 minutes there are 10 000.

a Show that $N = 1000e^{kt}$ and find the exact value of k in the form $a \ln b$.

b Calculate the number of bacteria after 8 minutes, giving your answer to two significant figures.

Note: This is the situation described in Example 4.10.

Step 1: Use the given information to form a differential equation.

a The rate of change of N is $\frac{dN}{dt}$.

$$\frac{dN}{dt} \propto N \Rightarrow \frac{dN}{dt} = kN, \text{ where } k \text{ is a positive constant}$$

Note: k is the proportionality constant.

Step 2: Separate the variables and integrate.

$$\frac{dN}{dt} = kN$$
$$\Rightarrow \int \frac{1}{N}\,dN = \int k\,dt$$
$$\Rightarrow \ln N = kt + c$$

Step 3: Substitute the first given condition to find c.

When $t = 0$, $N = 1000$.

So $\ln 1000 = k \times 0 + c$

$c = \ln 1000$

Hence $\ln N = kt + \ln 1000$

$\ln N - \ln 1000 = kt$

$$\ln \frac{N}{1000} = kt$$

$$\frac{N}{1000} = e^{kt}$$

$$N = 1000e^{kt}$$

Note:
Knowing the initial number of bacteria enables the integration constant to be found.

Tip:
As in Example 5.25 you could write $N = Ae^{kt}$ straight away and then find A.

Step 4: Substitute the second given condition to find k.

When $t = 5$, $N = 10\,000$.

So $10\,000 = 1000e^{k \times 5}$

$e^{5k} = 10$

$5k = \ln 10$

$k = \frac{1}{5}\ln 10$

Note:
Knowing a second condition enables the proportionality constant to be found.

Recall:
$e^b = a \Leftrightarrow \ln a = b$

Step 5: Write out the particular solution and substitute $t = 8$.

b $N = 1000e^{\frac{1}{5}\ln 10 \times t}$

When $t = 8$,

$$N = 1000e^{\frac{1}{5}\ln 10 \times 8}$$
$$= 1000e^{\frac{8}{5}\ln 10}$$
$$= 39\,810.7\ldots$$
$$= 40\,000 \text{ (2 s.f.)}$$

After 8 minutes, there are approximately 40 000 bacteria.

SKILLS CHECK **5F: Differential equations**

1 a Obtain the general solution of the differential equation $\frac{dy}{dx} = x(y + 1)$.

b Given that $y = 2$ when $x = 0$, find the particular solution, expressing y as a function of x.

2 a Obtain the general solution of the differential equation $\frac{dy}{dx} = \frac{1}{2}y^3x^2$.

b Given also that $y = 1$ at $x = 3$, show that $y^2 = \frac{3}{30 - x^3}$, $x \neq \sqrt[3]{30}$.

3 A curve has equation $y = f(x)$. The gradient of the curve at the point (x, y) is given by

$$\frac{dy}{dx} = \frac{2x}{1 - 2y}$$

Given also that $(0, 2.5)$ lies on the curve, show that the curve is a circle and find the coordinates of the centre of the circle, and its radius.

4 Given that $x = 1$ when $t = 1$, express x in terms of t, where

a $\frac{dx}{dt} = \frac{x + 3}{4t}$ **b** $\frac{dx}{dt} = \frac{x + 3}{4}$

5 Find the general solution of the differential equation $\frac{d\theta}{dx} = x\cos^2\theta$.

6 **a** Find **i** $\int x\cos 2x\,dx$ **ii** $\int \cos^2 x\,dx$

b Given that $y = 0$ at $x = \frac{1}{4}\pi$, solve the differential equation $\frac{dy}{dx} = \frac{x\cos 2x}{\cos^2 y}$.

7 **a** Using the substitution $u = 9 + x^3$, find $\int x^2(9 + x^3)^4\,dx$.

b Given that $y = 0$ when $x = -2$, solve the differential equation $\frac{dy}{dx} = \frac{15x^2(9 + x^3)^4}{e^y}$, expressing your answer in the form $y = \ln f(x)$.

8 According to Newton's law of cooling, the rate of temperature loss of a body is proportional to the difference between the temperature of the body, T, and the temperature of the surrounding air. The air in a room has a constant temperature of 15 °C.

a Show that $\frac{dT}{dt} = -k(T - 15)$, where k is a positive constant.

b Show, by integration, that $T = 15 + Ae^{-kt}$, where A is a constant.

The temperature of an object in the room is found to be 75 °C. Ten minutes later its temperature has dropped by 10 °C.

c Show that $k = 0.1 \ln 1.2$.

d Find the temperature of the object after a further ten minutes.

SKILLS CHECK **5F EXTRA is on the CD**

5.7 Numerical integration of functions

Application of the trapezium rule to functions covered in C3 and C4.

In *Core 2* you used the trapezium rule to find an approximate value for the area of the region bounded by the curve $y = f(x)$, the x-axis and the lines $x = a$ and $x = b$.

The region is split into n strips, each of width h, and trapezia are formed by joining the top ends of each strip with straight lines. The sum of the areas of these trapezia is calculated using the trapezium rule.

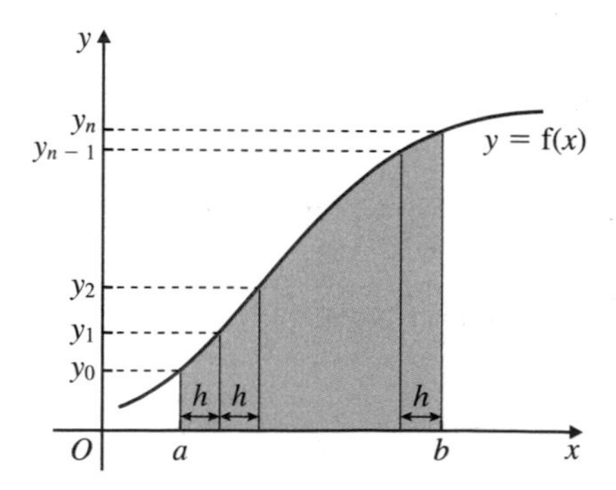

The value of the area given by the trapezium rule is used as an estimate of $\int_a^b y\,dx$.

The formula for the **trapezium rule** is

$$\int_a^b y\,dx \approx \tfrac{1}{2}h[(y_0 + y_n) + 2(y_1 + y_2 + \cdots + y_{n-1})] \quad \text{where } h = \frac{b - a}{n}$$

You will be expected to apply the trapezium rule to functions covered in *Core 3* and *Core 4*. You may also be asked to integrate to find the exact value of the area and to compare the two values.

Recall:
Trapezium rule (C2 Section 7.3).

Recall:
The y-value for each corresponding x-value is calculated. These y-values y_0, y_1, …, y_n are called **ordinates**.

Recall:
The accuracy of the estimate is improved by increasing the number of strips, thus reducing the width of each.

Note:
Some trapezia give an overestimate.

Some give an underestimate.

Recall:
Make sure you know where to find the formula in the formulae booklet and that you transfer it accurately.

Example 5.27 The diagram shows the curve with equation

$$y = xe^{-3x}, \quad x \geqslant 0$$

The finite region R, bounded by the x-axis, the line $x = 1$ and the curve, is shown shaded.

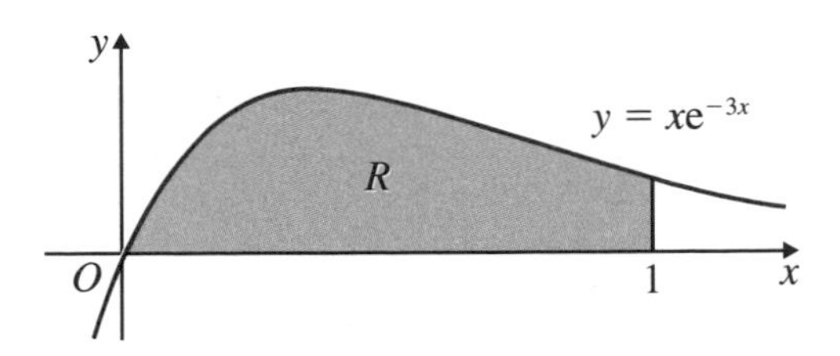

a Use the trapezium rule, with six ordinates, to find an approximate value for the area of R, giving your answer to four significant figures.

b Use integration by parts to show that the exact value of the area of R is $\frac{1}{9}(1 - 4e^{-3})$.

c **i** Calculate the percentage error in taking your answer in part **a** for the area of R.

ii Explain how you could increase the accuracy of your estimate, using the trapezium rule.

Step 1: Find h.

a 6 ordinates $\Rightarrow$ 5 strips

$$n = 5, \quad a = 0, \quad b = 1, \quad h = \frac{b - a}{n} = \frac{1 - 0}{5} = 0.2$$

Step 2: Substitute values into the trapezium rule.

	End ordinates	**Middle ordinates**
$x_0 = 0$	$y_0 = 0$	
$x_1 = 0.2$		$y_1 = 0.2e^{-0.6}$
$x_2 = 0.4$		$y_2 = 0.4e^{-1.2}$
$x_3 = 0.6$		$y_3 = 0.6e^{-1.8}$
$x_4 = 0.8$		$y_4 = 0.8e^{-2.4}$
$x_5 = 1$	$y_5 = e^{-3}$	

Tip: You need the sum of the end ordinates and twice the sum of the middle ordinates.

$$\int_a^b y \, dx \approx \tfrac{1}{2} h \, [(y_0 + y_5) + 2(y_1 + y_2 + \cdots + y_4)]$$

So $\text{area} \approx \tfrac{1}{2} \times 0.2 \, [(0 + e^{-3})$

$$+ 2(0.2e^{-0.6} + 0.4e^{-1.2} + 0.6e^{-1.8} + 0.8e^{-2.4})]$$

$$= 0.1 \times 0.853\,77\ldots$$

$$= 0.085\,377\ldots$$

$$= 0.085\,38 \text{ (4 s.f.)}$$

Tip: To avoid rounding errors, do not calculate the values until the final stage of working.

Step 1: Write out the area formula using integration.

b Area $= \int_0^1 x\mathrm{e}^{-3x}\,\mathrm{d}x$

Using integration by parts:

Step 2: Decide on the functions u and $\frac{\mathrm{d}v}{\mathrm{d}x}$, and find $\frac{\mathrm{d}u}{\mathrm{d}x}$ and v.

$$\int_a^b u\frac{\mathrm{d}v}{\mathrm{d}x}\,\mathrm{d}x = \Big[uv\Big]_a^b - \int_a^b v\frac{\mathrm{d}u}{\mathrm{d}x}\,\mathrm{d}x$$

Let $u = x$, then $\frac{\mathrm{d}u}{\mathrm{d}x} = 1$

Let $\frac{\mathrm{d}v}{\mathrm{d}x} = \mathrm{e}^{-3x}$, then $v = \int \mathrm{e}^{-3x}\,\mathrm{d}x = -\frac{1}{3}\mathrm{e}^{-3x}$

Step 3: Apply the formula for integrating by parts.

$$\int_0^1 x\mathrm{e}^{-3x}\,\mathrm{d}x = \Big[x \times (-\tfrac{1}{3}\mathrm{e}^{-3x})\Big]_0^1 - \int_0^1 -\tfrac{1}{3}\mathrm{e}^{-3x} \times 1\,\mathrm{d}x$$

$$= -\tfrac{1}{3}\Big[x\mathrm{e}^{-3x}\Big]_0^1 + \int_0^1 \tfrac{1}{3}\mathrm{e}^{-3x}\,\mathrm{d}x$$

$$= -\tfrac{1}{3}(\mathrm{e}^{-3} - 0) + \Big[\tfrac{1}{3} \times (-\tfrac{1}{3}\mathrm{e}^{-3x})\Big]_0^1$$

$$= -\tfrac{1}{3}\mathrm{e}^{-3} - \tfrac{1}{9}\Big[\mathrm{e}^{-3x}\Big]_0^1$$

$$= -\tfrac{1}{3}\mathrm{e}^{-3} - \tfrac{1}{9}(\mathrm{e}^{-3} - \mathrm{e}^{0})$$

$$= -\tfrac{1}{3}\mathrm{e}^{-3} - \tfrac{1}{9}\mathrm{e}^{-3} + \tfrac{1}{9}$$

$$= \tfrac{1}{9} - \tfrac{4}{9}\mathrm{e}^{-3}$$

$$= \tfrac{1}{9}(1 - 4\mathrm{e}^{-3})$$

Recall:

Integration by parts (Section 5.4).

Recall:

$\int \mathrm{e}^{kx}\,\mathrm{d}x = \frac{1}{k}\mathrm{e}^{kx}$ (Section 5.1).

Tip:

You may prefer to find the indefinite integral first, then substitute the limits at the end.

Step 1: Calculate the error and express it as a percentage of the true value.

c i Error $= \frac{1}{9}(1 - 4\mathrm{e}^{-3}) - 0.085\,377\ldots$

Percentage error $= \frac{\text{error}}{\text{true value}} \times 100$

$$= \frac{\frac{1}{9}(1 - 4\mathrm{e}^{-3}) - 0.085\,377\ldots}{\frac{1}{9}(1 - 4\mathrm{e}^{-3})} \times 100$$

$= 4.049\ldots$

The percentage error is 4.0% (2 s.f.).

Step 2: State an appropriate method.

ii To increase the accuracy using the trapezium rule, increase the number of strips.

SKILLS CHECK **5G: Numerical integration of functions**

1 Use the trapezium rule with four strips to estimate $\int_1^3 \sqrt{\mathrm{e}^x + 1}\,\mathrm{d}x$, giving your answer to three significant figures.

2 Use the trapezium rule with five strips to estimate $\int_0^1 \mathrm{e}^{\cos x}\,\mathrm{d}x$, giving your answer to three significant figures.

3 a Use the trapezium rule with four strips to find an approximate value for $\int_2^6 \sqrt{\ln(x^2 - 1)}\,\mathrm{d}x$.

b Explain briefly how the trapezium rule could be used to find a more accurate estimate of the integral.

4 The following is a table of values, correct to three decimal places, for $y = \cos^2 x$, where x is in radians.

x	0	0.2	0.4	0.6	0.8	1
y	1	0.961	a	0.681	b	0.292

a Find the value of a and the value of b.

b Use the trapezium rule and the values of y in the completed table to obtain an estimate for $\int_0^1 \cos^2 x \, dx$.

c Using integration, calculate the exact value of $\int_0^1 \cos^2 x \, dx$.

d Using your answers from parts **b** and **c**, calculate the percentage error of your estimate.

5 The diagram shows the curve

$$y = \frac{1}{\sqrt{\ln (x + 1)}}.$$

The region enclosed by the curve, the x-axis and the lines $x = 1$ and $x = 4$, shown shaded, is rotated completely about the x-axis.

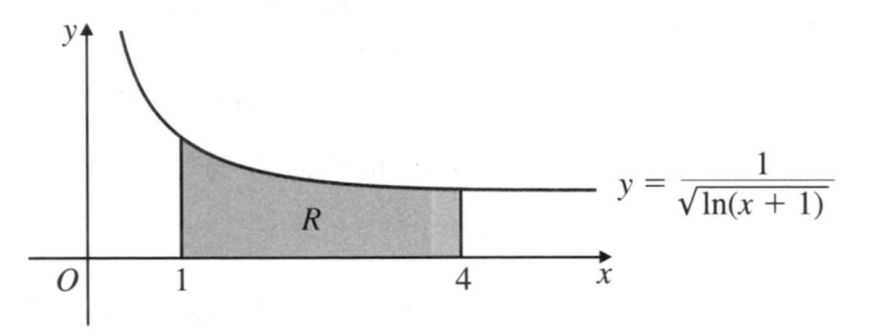

Use the trapezium rule with six strips to estimate the value of the volume of the solid formed.

SKILLS CHECK **5G EXTRA is on the CD**

Examination practice 5: Integration

1 The finite region R is bounded by the curve $y = e^{-x}$, the coordinate axes and the line $x = -1$.

a Find the exact value of the area of the region R.

The region R is rotated completely about the x-axis.

b Find the exact value of the volume of the solid generated.

2 Given that $y = 0$ at $x = 1$, solve the differential equation

$$\frac{dy}{dx} = e^{x+y}$$

giving your answer exactly in the form $y = f(x)$.

3 **a** **i** Express $\sin^2 A$ in terms of $\cos 2A$. **ii** Find $\int \sin^2 x \, dx$.

b Use the substitution $x = \sin \theta$ to find the exact value of $\int_0^{\frac{1}{2}} \frac{x^2}{\sqrt{1 - x^2}} \, dx$.

4 $f(x) \equiv \frac{5x^2 - 8x + 1}{2x(x - 1)^2} \equiv \frac{A}{x} + \frac{B}{x - 1} + \frac{C}{(x - 1)^2}$.

a Find the values of the constant A, B and C.

b Hence find $\int f(x) \, dx$.

c Hence show that $\int_4^9 f(x) \, dx = \ln (\frac{32}{3}) - \frac{5}{24}$. [London Jan 1998]

5 **a** Using the substitution $u = 1 + 2x^2$, find $\int x(1 + 2x^2)^5 \, dx$.

b Given that $y = \frac{\pi}{8}$ at $x = 0$, solve the differential equation

$$\frac{dy}{dx} = x(1 + 2x^2)^5 \cos^2 2y.$$

[Edexcel June 2000]

6 **a** Use integration by parts to find $\int x \cos 2x \, dx$.

b Prove that the answer to part **a** may be expressed as

$\frac{1}{2} \sin x \,(2x \cos x - \sin x) + C,$

where C is an arbitrary constant.

[Edexcel May 2002]

7 The diagram shows a sketch of the curve C with parametric equations

$$x = 3t \sin t, \quad y = 2 \sec t, \qquad 0 \leqslant t < \frac{\pi}{2}.$$

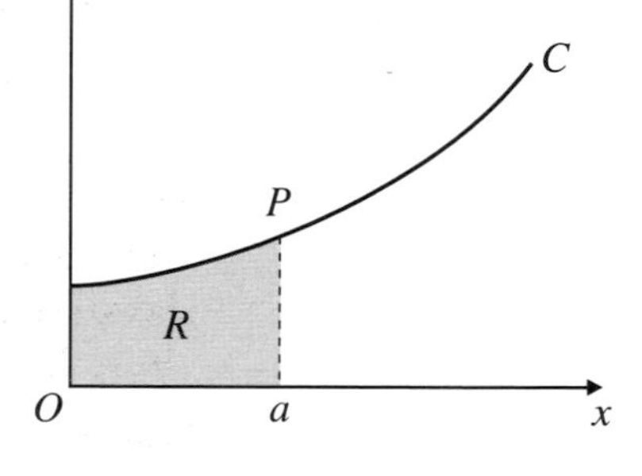

The point $P(a, 4)$ lies on C.

a Find the exact value of a.

The region R is enclosed by C, the axes and the line $x = a$ as shown in the diagram.

b Show that the area of R is given by

$$6\int_0^{\frac{\pi}{3}} (\tan t + t) \, dt.$$

c Find the exact value of the area of R.

[Edexcel June 2004]

8 Use the substitution $x = \tan \theta$ to show that

$$\int_0^1 \frac{1}{(1 + x^2)^2} \, dx = \frac{\pi}{8} + \frac{1}{4}.$$

[Edexcel C4 Specimen]

9 Use the substitution $u = 4 + 3x^2$ to find the exact value of

$$\int_0^2 \frac{2x}{(4 + 3x^2)^2} \, dx.$$

[Edexcel C4 Mock]

10 $\frac{dy}{dx} = \frac{2xy}{1 + x^2}, \quad y > 0.$

a Find the general solution of this differential equation.

Given that $y = 2$ at $x = 0$,

b find the particular solution in the form $y = f(x)$.

[London May 1995]

11 **a** **i** Find $\int \frac{1}{x(x + 1)} \, dx, \; x > 0.$

Using the substitution $u = e^x$ and the answer to **i**, or otherwise,

ii find $\int \frac{1}{1 + e^x} \, dx.$

b Use integration by parts to find $\int x^2 \sin x \, dx$.

[London June 1997]

12 a Given that

$$\frac{x^2}{x^2-1} \equiv A + \frac{B}{x-1} + \frac{C}{x+1},$$

find the values of the constants A, B and C.

b Given that $x = 2$ at $t = 1$, solve the differential equation

$$\frac{dx}{dt} = 2 - \frac{2}{x^2}, x > 1.$$

You need not simplify your final answer. [Edexcel Jan 2001]

13 The diagram shows part of the curve with equation

$$y = e^x \cos x, \quad 0 \leqslant x \leqslant \frac{\pi}{2}.$$

The finite region R is bounded by the curve and the coordinate axes.

a Calculate, to 2 decimal places, the y-coordinates of the points on the curve where $x = 0, \frac{\pi}{6}, \frac{\pi}{3}$ and $\frac{\pi}{2}$.

b Using the trapezium rule and all the values calculated in part **a**, find an approximation for the area of R.

c State, with a reason, whether your approximation underestimates or overestimates the area of R. [Edexcel Nov 2003]

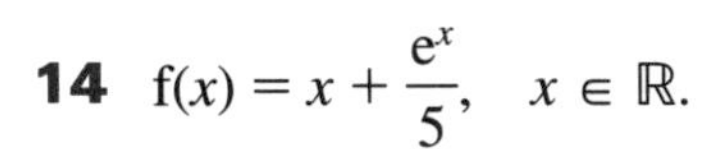

14 $f(x) = x + \frac{e^x}{5}, \quad x \in \mathbb{R}.$

a Find $f'(x)$.

The curve C, with equation $y = f(x)$, crosses the y-axis at the point A.

b Find an equation for the tangent to C at A.

c Complete the table, giving the values of $\sqrt{\left(x + \frac{e^x}{5}\right)}$ to 2 decimal places.

x	0	0.5	1	1.5	2
$\sqrt{\left(x + \frac{e^x}{5}\right)}$	0.45	0.91			

d Use the trapezium rule, with all the values from your table, to find an approximation for the value of

$$\int_0^2 \sqrt{\left(x + \frac{e^x}{5}\right)}\, dx.$$ [Edexcel June 2004]

15 The diagram shows part of the curve with equation

$$y = 4x - \frac{6}{x}, \quad x > 0.$$

The shaded region R is bounded by the curve, the x-axis and the lines with equations $x = 2$ and $x = 4$. This region is rotated through 2π radians about the x-axis.

Find the exact value of the volume of the solid generated. [Edexcel Nov 2004]

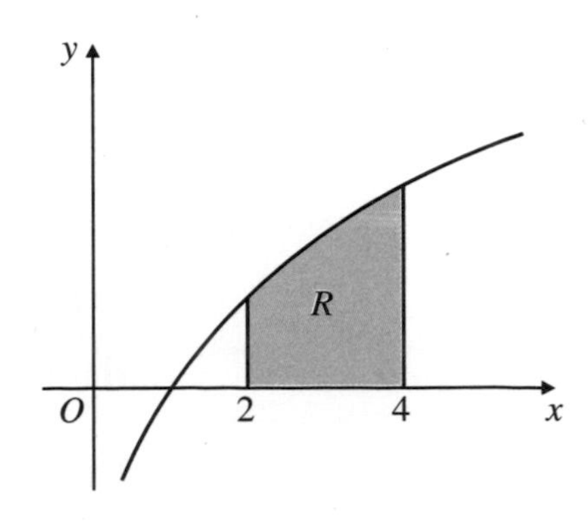

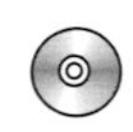 **16** The curve C with equation $y = k + \ln 2x$, where k is a constant, crosses the x-axis at the point $A\left(\frac{1}{2e}, 0\right)$.

a Show that $k = 1$.

b Show that an equation of the tangent to C at A is $y = 2ex - 1$.

c Complete the table below, giving your answers to 3 significant figures.

x	1	1.5	2	2.5	3
$\mathbf{1 + \ln 2x}$		2.10		2.61	2.79

d Use the trapezium rule, with four equal intervals, to estimate the value of

$$\int_1^3 (1 + \ln 2x)\,dx.$$

[Edexcel Nov 2004]

6 Vectors

6.1 Vector geometry

Vectors in two and three dimensions. Magnitude of a vector. Algebraic operations of vector addition and multiplication by scalars, and their geometrical interpretations.

A **scalar** is a quantity that can be expressed by magnitude (size) alone, for example length, distance, speed, volume.

A **vector** is a quantity that it is expressed in terms of magnitude and direction, for example displacement, velocity, acceleration, momentum.

So, as an example, wind speed is a scalar quantity and can be expressed in terms of its magnitude, such as 50 km/h. However, wind velocity is a vector quantity and therefore needs a direction given as well, for example 50 km/h from the south-west.

Vector notation

A directed line segment is drawn to represent a vector. The length of the line represents the magnitude of the vector and an arrow is used to represent the direction of the vector.

Note:
A directed line segment is simply a straight line with an arrow to show direction.

For example, the vector of the displacement from P to Q can be represented as follows:

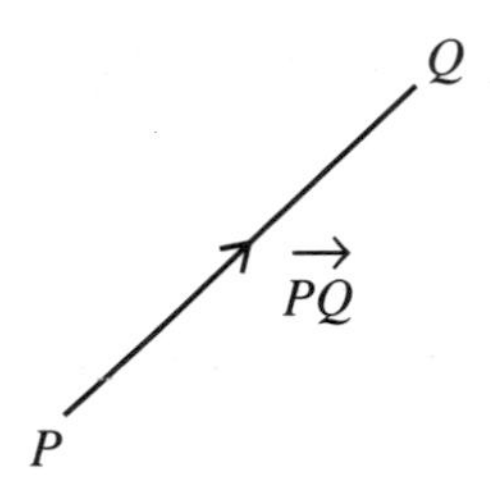

or, simply,

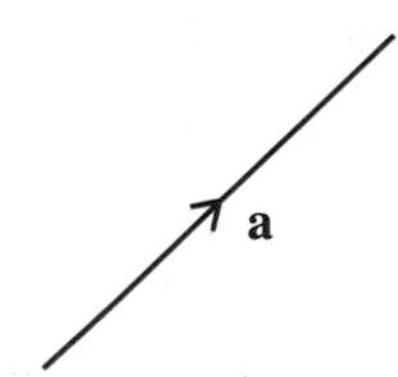

Note:
In print, a lower case letter will be identified as a vector by the use of **bold** type. In your work you must use an underlined lower case letter, $\underline{a}$.

where $\overrightarrow{PQ} = \mathbf{a}$

Vectors are **equal** if they have the same magnitude and the same direction.

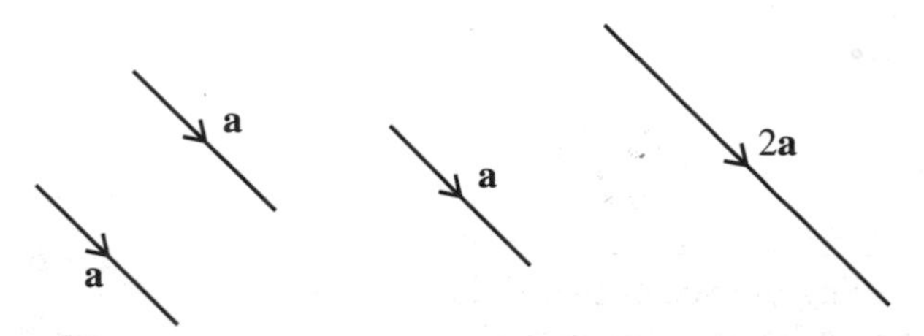

Note:
Each of the three lines on the left of the diagram represents the same vector, **a**, because the lines are all of the same length and direction. The vector on the right, 2**a**, is a vector in the same direction as **a**, but with twice the magnitude.

Any vector **parallel** to a vector **a** can be written in the form $\lambda\mathbf{a}$, where λ is a scalar multiple. The vector $\lambda\mathbf{a}$ will have magnitude λ times the magnitude of **a**.

Recall:
A scalar is a number.

Now, if P is the point (1, 2) and Q is the point (4, 6), then the vector $\overrightarrow{PQ} = \begin{pmatrix} 3 \\ 4 \end{pmatrix}$, where 3 is the increase in the x-coordinate and 4 is the increase in the y-coordinate.

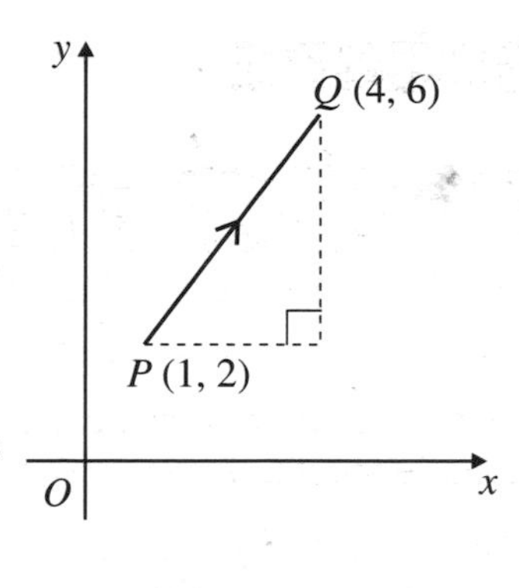

Tip: This means that to get from P to Q you go 3 units right and 4 units up.

Tip: Vectors are written as column vectors, e.g. $\begin{pmatrix} 3 \\ 4 \end{pmatrix}$; don't confuse these with coordinates.

From the diagram, the vector with the same magnitude as $\overrightarrow{PQ}$, but in the opposite direction, is $\overrightarrow{QP} = \begin{pmatrix} -3 \\ -4 \end{pmatrix} = -\overrightarrow{PQ}$. This is true for all vectors.

Tip: To get from Q to P you go 3 units left and 4 units down.

The **magnitude** of the vector $\overrightarrow{PQ}$ is also called its modulus and is written $|\overrightarrow{PQ}|$.

Note: If you are using lower case letters for the vector e.g. **a**, then the magnitude is written $|\mathbf{a}|$.

In the above example, the length of $\overrightarrow{PQ}$, i.e. $|\overrightarrow{PQ}|$, can be found using Pythagoras' theorem:

$$|\overrightarrow{PQ}|^2 = 3^2 + 4^2 = 9 + 16 = 25$$
$$|\overrightarrow{PQ}| = \sqrt{25} = 5$$

Tip: If you notice it is a 3, 4, 5 triangle you could use this fact to save having to work it out.

A **unit vector** has magnitude 1. To find a unit vector in the direction of the vector **a**, divide **a** by its magnitude, i.e. find $\dfrac{\mathbf{a}}{|\mathbf{a}|}$.

In the above example, a unit vector parallel to $\overrightarrow{PQ}$ is $\dfrac{\overrightarrow{PQ}}{|\overrightarrow{PQ}|} = \frac{1}{5}\overrightarrow{PQ}$.

Addition of vectors

To add vectors use the **triangle law** for vector addition.

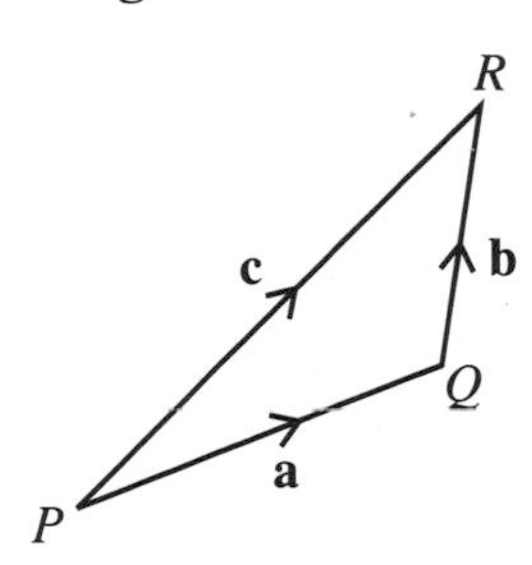

Tip: Trace your finger along **a** then **b** in the direction of the arrows. You are starting at P and ending at R, which is the same outcome as tracing your finger along **c**.

Thinking of $\overrightarrow{PQ}$ as a displacement vector equivalent to travelling from P to Q, and similarly $\overrightarrow{QR}$ as a journey from Q to R, then $\overrightarrow{PQ} + \overrightarrow{QR}$ is the **resultant** journey from P to R, $\overrightarrow{PR}$, i.e.

$$\overrightarrow{PR} = \overrightarrow{PQ} + \overrightarrow{QR}$$

or $\mathbf{c} = \mathbf{a} + \mathbf{b}$

Subtraction of vectors

To work out $\mathbf{a} - \mathbf{b}$ think of it as $\mathbf{a} + (-\mathbf{b})$, so that you are adding the vector that is the same magnitude as **b** but is in the opposite direction.

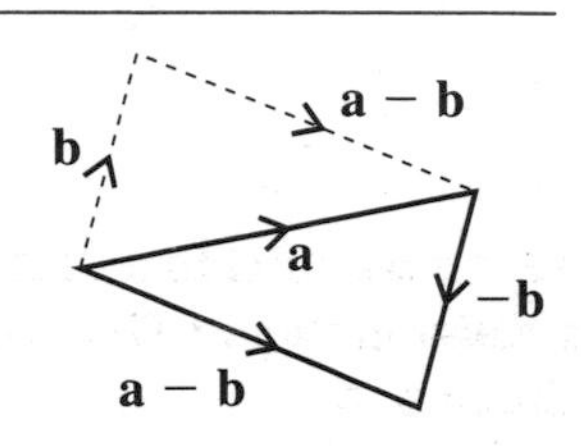

Tip: Tracing your finger backwards along **b**, then along **a** means that you move the same distance and direction as tracing along **a** and then $-$**b**.

Example 6.1 In the diagram, $\overrightarrow{AB} = \mathbf{p}$, $\overrightarrow{AF} = \mathbf{q}$, and $\overrightarrow{BF} = \overrightarrow{FC}$.

a Find, in terms of **p** and **q**,

i $\overrightarrow{BF}$

ii $\overrightarrow{BC}$

iii $\overrightarrow{AC}$

b D and E are the midpoints of AB and AF.

i Find $\overrightarrow{DE}$ in terms of **p** and **q**.

ii Prove that DE is parallel to BF.

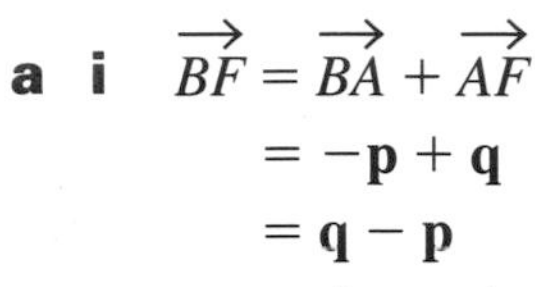

Step 1: Use the triangle law to add vectors.

a i $\overrightarrow{BF} = \overrightarrow{BA} + \overrightarrow{AF}$

$= -\mathbf{p} + \mathbf{q}$

$= \mathbf{q} - \mathbf{p}$

Tip: Use $\triangle ABF$: $\overrightarrow{BA} = -\overrightarrow{AB} = -\mathbf{p}$

Step 2: Use the given relation with your answer from **a**.

ii $\overrightarrow{BF} = \overrightarrow{FC}$

so $\overrightarrow{BC} = 2\overrightarrow{BF}$

$\overrightarrow{BC} = 2(\mathbf{q} - \mathbf{p})$

Step 3: Use the triangle law to add vectors.

iii $\overrightarrow{AC} = \overrightarrow{AB} + \overrightarrow{BC}$

$= \mathbf{p} + 2(\mathbf{q} - \mathbf{p})$

$= \mathbf{p} + 2\mathbf{q} - 2\mathbf{p} = 2\mathbf{q} - \mathbf{p}$

Tip: Use $\triangle ABC$.

Step 1: Use the given relations and the triangle law to add vectors.

b i $\overrightarrow{AD} = \frac{1}{2}\overrightarrow{AB} = \frac{1}{2}\mathbf{p}$

$\overrightarrow{AE} = \frac{1}{2}\overrightarrow{AF} = \frac{1}{2}\mathbf{q}$

$\overrightarrow{DE} = \overrightarrow{DA} + \overrightarrow{AE}$

$= -\frac{1}{2}\mathbf{p} + \frac{1}{2}\mathbf{q}$

$= \frac{1}{2}\mathbf{q} - \frac{1}{2}\mathbf{p}$

Tip: In **b i**, if D is the midpoint of AB, then the vector $\overrightarrow{AD}$ must be in the same direction as $\overrightarrow{AB}$ and half its magnitude.

Tip: Use $\triangle ADE$: $\overrightarrow{DA} = -\overrightarrow{AD} = -\frac{1}{2}\mathbf{p}$

Step 2: Write $\overrightarrow{DE}$ in terms of $\overrightarrow{BF}$ and state your conclusion.

ii $\overrightarrow{DE} = \frac{1}{2}\mathbf{q} - \frac{1}{2}\mathbf{p} = \frac{1}{2}(\mathbf{q} - \mathbf{p}) = \frac{1}{2}\overrightarrow{BF}$

Hence DE is parallel to BF.

Recall: Two vectors are parallel if one is a scalar multiple of the other.

Vectors in two and three dimensions

A **unit vector** in the direction of the x-axis is called **i** and represents 1 unit across (in the positive x-direction):

$$\mathbf{i} = \begin{pmatrix} 1 \\ 0 \end{pmatrix}$$

A unit vector in the direction of the y-axis is called **j** and represents 1 unit up (in the y-direction):

$$\mathbf{j} = \begin{pmatrix} 0 \\ 1 \end{pmatrix}$$

Any vector in two dimensions can be written in terms of **i** and **j**.

For example, the vector $\overrightarrow{PQ} = \begin{pmatrix} 3 \\ 4 \end{pmatrix}$ can be written as $\overrightarrow{PQ} = 3\mathbf{i} + 4\mathbf{j}$.

Similarly, in three dimensions the unit vectors $\mathbf{i} = \begin{pmatrix} 1 \\ 0 \\ 0 \end{pmatrix}$, $\mathbf{j} = \begin{pmatrix} 0 \\ 1 \\ 0 \end{pmatrix}$ and $\mathbf{k} = \begin{pmatrix} 0 \\ 0 \\ 1 \end{pmatrix}$ can be used.

Note: Here **i** represents 1 unit in the x-direction, **j** represents 1 unit in the y-direction and **k** represents 1 unit in the z-direction.

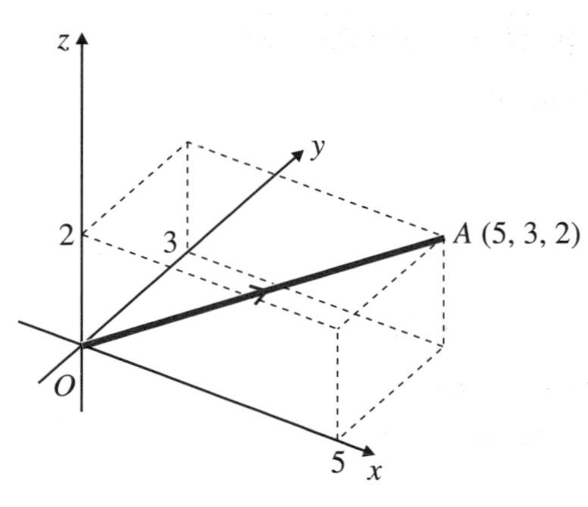

Tip: Imagine the x- and y-axes flat on the table, then the z-axis rises perpendicularly upwards.

Consider the vector that joins the points $O(0, 0, 0)$ and $A(5, 3, 2)$.

$$\overrightarrow{OA} = \begin{pmatrix} 5 \\ 3 \\ 2 \end{pmatrix} = 5\mathbf{i} + 3\mathbf{j} + 2\mathbf{k}$$

Note: $\overrightarrow{OA}$ is a position vector (see Section 6.2).

As before, to find the magnitude or modulus of a vector, use Pythagoras' theorem.

- The magnitude of the vector $\mathbf{v} = x\mathbf{i} + y\mathbf{j}$ is given by
  $|\mathbf{v}| = \sqrt{x^2 + y^2}$
- The magnitude of the vector $\mathbf{v} = x\mathbf{i} + y\mathbf{j} + z\mathbf{k}$ is given by
  $|\mathbf{v}| = \sqrt{x^2 + y^2 + z^2}$

Note: This is Pythagoras' theorem in three dimensions.

Example 6.2 Given the vectors $\mathbf{p} = 4\mathbf{i} + 2\mathbf{j} - 5\mathbf{k}$ and $\mathbf{q} = 2\mathbf{i} - \mathbf{j} + 2\mathbf{k}$, find

a $\mathbf{p} + 2\mathbf{q}$,

b the exact value of $|\mathbf{p} - \mathbf{q}|$,

c a unit vector in the direction of $\mathbf{p}$.

Step 1: Add the vectors by collecting the coefficients of **i**, **j** and **k**.

a $\mathbf{p} + 2\mathbf{q} = (4\mathbf{i} + 2\mathbf{j} - 5\mathbf{k}) + 2(2\mathbf{i} - \mathbf{j} + 2\mathbf{k})$
$= (4\mathbf{i} + 2\mathbf{j} - 5\mathbf{k}) + (4\mathbf{i} - 2\mathbf{j} + 4\mathbf{k})$
$= 8\mathbf{i} - \mathbf{k}$

Tip: The brackets make your working clearer.

Tip: $2\mathbf{j} - 2\mathbf{j} = 0\mathbf{j}$

Step 1: Find $\mathbf{p} - \mathbf{q}$.

b $\mathbf{p} - \mathbf{q} = (4\mathbf{i} + 2\mathbf{j} - 5\mathbf{k}) - (2\mathbf{i} - \mathbf{j} + 2\mathbf{k})$
$= 2\mathbf{i} + 3\mathbf{j} - 7\mathbf{k}$

Step 2: Use Pythagoras' theorem to find the modulus.

$|\mathbf{p} - \mathbf{q}| = \sqrt{2^2 + 3^3 + (-7)^2}$
$= \sqrt{62}$

Tip: Be careful with the signs.

Note: $|\mathbf{p} - \mathbf{q}| \neq |\mathbf{p}| - |\mathbf{q}|$

Tip: Part **b** asks for the exact value, so leave your answer in surd form.

Step 1: Use Pythagoras' theorem to find the modulus.
Step 2: Divide the vector by its modulus.

c $\mathbf{p} = 4\mathbf{i} + 2\mathbf{j} - 5\mathbf{k}$
$|\mathbf{p}| = \sqrt{4^2 + 2^2 + (-5)^2} = \sqrt{45} = 3\sqrt{5}$

Unit vector is $\dfrac{\mathbf{p}}{|\mathbf{p}|} = \dfrac{4\mathbf{i} + 2\mathbf{j} - 5\mathbf{k}}{3\sqrt{5}}$

$$= \frac{4}{3\sqrt{5}}\mathbf{i} + \frac{2}{3\sqrt{5}}\mathbf{j} - \frac{5}{3\sqrt{5}}\mathbf{k}$$

Note: You could tidy this up to $\dfrac{4\sqrt{5}}{15}\mathbf{i} + \dfrac{2\sqrt{5}}{15}\mathbf{j} - \dfrac{\sqrt{5}}{3}\mathbf{k}$

Example 6.3 Given the vector $\mathbf{a} = 3\mathbf{i} - \mathbf{j} + 2\mathbf{k}$, state whether each of the following vectors is equal to **a**, parallel to **a**, or neither.

a $-6\mathbf{i} + 2\mathbf{j} - 4\mathbf{k}$ **b** $6\mathbf{i} - 2\mathbf{j} - 4\mathbf{k}$ **c** $-\frac{1}{3}(-9\mathbf{i} + 3\mathbf{j} - 6\mathbf{k})$

Step 1: Check whether the vector is a scalar multiple of **a** or equal to **a**.

a $-6\mathbf{i} + 2\mathbf{j} - 4\mathbf{k} = -2(3\mathbf{i} - \mathbf{j} + 2\mathbf{k})$
Therefore $-6\mathbf{i} + 2\mathbf{j} - 4\mathbf{k}$ is parallel to **a**.

Tip: The scalar multiple is -2.

b $6\mathbf{i} - 2\mathbf{j} - 4\mathbf{k} = 2(3\mathbf{i} - \mathbf{j} - 2\mathbf{k})$ so it is not a multiple of **a**.
Therefore $6\mathbf{i} - 2\mathbf{j} - 4\mathbf{k}$ is neither parallel nor equal to **a**.

c $-\frac{1}{3}(-9\mathbf{i} + 3\mathbf{j} - 6\mathbf{k}) = -\frac{1}{3}(-3)(3\mathbf{i} - \mathbf{j} + 2\mathbf{k}) = 3\mathbf{i} - \mathbf{j} + 2\mathbf{k}$
Therefore $-\frac{1}{3}(-9\mathbf{i} + 3\mathbf{j} - 6\mathbf{k})$ is equal to $3\mathbf{i} - \mathbf{j} + 2\mathbf{k}$.

Example 6.4 Given the vectors $\overrightarrow{PQ} = 4\mathbf{i} + \mathbf{j} - 2\mathbf{k}$ and $\overrightarrow{RQ} = 2\mathbf{i} - \mathbf{j} - \mathbf{k}$, find $\overrightarrow{PR}$.

Step 1: Draw a sketch to represent the vectors.

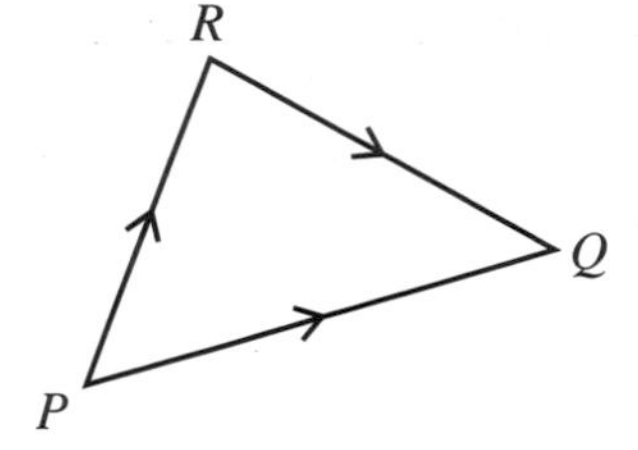

Note: The diagram is not meant to be an accurate representation, just a sketch to help.

Step 2: Use the triangle law to add the vectors.

$$\overrightarrow{PR} = \overrightarrow{PQ} + \overrightarrow{QR}$$
$$= (4\mathbf{i} + \mathbf{j} - 2\mathbf{k}) - (2\mathbf{i} - \mathbf{j} - \mathbf{k})$$
$$= 2\mathbf{i} + 2\mathbf{j} - \mathbf{k}$$

Tip: $\overrightarrow{QR} = -\overrightarrow{RQ}$

If you prefer, you can work in terms of column vectors.

In the above example $\overrightarrow{PQ} = \begin{pmatrix} 4 \\ 1 \\ -2 \end{pmatrix}$ and $\overrightarrow{RQ} = \begin{pmatrix} 2 \\ -1 \\ -1 \end{pmatrix}$.

So $\overrightarrow{PR} = \begin{pmatrix} 4 \\ 1 \\ -2 \end{pmatrix} - \begin{pmatrix} 2 \\ -1 \\ -1 \end{pmatrix} = \begin{pmatrix} 2 \\ 2 \\ -1 \end{pmatrix} = 2\mathbf{i} + 2\mathbf{j} - \mathbf{k}$

SKILLS CHECK **6A: Vector geometry**

1 Given the vector $\mathbf{a} = 2\mathbf{i} - \mathbf{j} + \mathbf{k}$, state whether each of the following vectors is equal to **a**, parallel to **a**, or neither.

a $-4\mathbf{i} + 2\mathbf{j} + 2\mathbf{k}$ **b** $\frac{1}{2}(4\mathbf{i} - 2\mathbf{j} + 2\mathbf{k})$ **c** $-8\mathbf{i} + 4\mathbf{j} - 4\mathbf{k}$

2 Given the vectors $\overrightarrow{AC} = 3\mathbf{i} + \mathbf{j} - \mathbf{k}$ and $\overrightarrow{BC} = -\mathbf{i} - 4\mathbf{j} + 5\mathbf{k}$, find $\overrightarrow{AB}$.

3 Find the magnitude of each of the following vectors.

a $5\mathbf{i} + 12\mathbf{j}$ **b** $2\mathbf{i} - 3\mathbf{j} + \mathbf{k}$ **c** $-3\mathbf{i} - 4\mathbf{k}$

d $\begin{pmatrix} 2 \\ 4 \end{pmatrix}$ **e** $\begin{pmatrix} 5 \\ -1 \\ 2 \end{pmatrix}$ **f** $\begin{pmatrix} 1 \\ -2 \\ -1 \end{pmatrix}$

4 Given that $\mathbf{a} = \mathbf{i} - 3\mathbf{j} + \mathbf{k}$, $\mathbf{b} = 3\mathbf{i} - \mathbf{j} - 4\mathbf{k}$ and $\mathbf{c} = 2\mathbf{i} + \mathbf{j} - \mathbf{k}$, find

a $\mathbf{a} + 2\mathbf{b}$ **b** $2\mathbf{c} - \mathbf{a}$ **c** $\mathbf{a} - \mathbf{b} + 3\mathbf{c}$

d $|\mathbf{b} - \mathbf{a}|$ **e** $|\mathbf{a} + \mathbf{b} + \mathbf{c}|$ **f** $|2\mathbf{a} - \mathbf{b} + 2\mathbf{c}|$

5 Given that $\mathbf{a} = \begin{pmatrix} 2 \\ 0 \\ -1 \end{pmatrix}$, $\mathbf{b} = \begin{pmatrix} -1 \\ -2 \\ -1 \end{pmatrix}$ and $\mathbf{c} = \begin{pmatrix} 3 \\ 3 \\ 3 \end{pmatrix}$, find

a $\mathbf{b} - 2\mathbf{c}$ **b** $\mathbf{a} - \mathbf{b} + \mathbf{c}$ **c** $3\mathbf{a} - \mathbf{b}$

d $|\mathbf{a} + \mathbf{c}|$ **e** $|2\mathbf{c} - \mathbf{a} - \mathbf{b}|$ **f** $|2\mathbf{a} + 2\mathbf{b} - \mathbf{c}|$

6 Find a unit vector in the direction of $\mathbf{i} + 5\mathbf{j} - 7\mathbf{k}$.

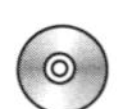

7 Given the points $A(-1, 1)$, $B(1, -2)$ and $C(2, 3)$,

a find the vectors

i $\overrightarrow{AB}$ **ii** $\overrightarrow{AC}$ **iii** $\overrightarrow{BC}$

b Find the magnitude of each of the vectors in part **a**.

c Hence prove that ABC is a right-angled triangle.

8 Vectors **a** and **b** are such that $\mathbf{a} = 3\mathbf{i} + 2\mathbf{j} + \mathbf{k}$, $\mathbf{b} = -\mathbf{i} + 6\mathbf{j} + \lambda\mathbf{k}$ and $|\mathbf{a} - \mathbf{b}| = 6$. Find the two possible values of λ.

SKILLS CHECK **6A EXTRA is on the CD**

6.2 Position vectors and distance

Position vectors. The distance between two points.

If you have a fixed origin, O, and a point, A, then the vector $\overrightarrow{OA} = \mathbf{a}$ is called the **position vector** of A.

Tip: As before, be careful not to confuse the coordinates of A (written horizontally) with the column vector, which represents the displacement from the origin to A.

So, for example, consider the point $A(2, 1, -1)$. A has position vector

$$\mathbf{a} = \overrightarrow{OA} = \begin{pmatrix} 2 \\ 1 \\ -1 \end{pmatrix} = 2\mathbf{i} + \mathbf{j} - \mathbf{k}$$

If the two points A and B have position vectors $\overrightarrow{OA} = \mathbf{a}$ and $\overrightarrow{OB} = \mathbf{b}$ respectively, then, using the triangle law,

Recall: Triangle law for vector addition (Section 6.1).

$$\begin{aligned} \overrightarrow{AB} &= \overrightarrow{AO} + \overrightarrow{OB} \\ &= -\overrightarrow{OA} + \overrightarrow{OB} \\ &= -\mathbf{a} + \mathbf{b} \\ &= \mathbf{b} - \mathbf{a} \end{aligned}$$

Example 6.5 Given the points A and B with position vectors **a** and **b** respectively, find the position vector of the midpoint of the line AB.

Tip: If M is the midpoint of AB, then the vector $\overrightarrow{AM}$ is half the vector $\overrightarrow{AB}$.

Let M be the midpoint of AB.

Step 1: Draw a diagram.

Step 2: Use the triangle law to add vectors.

Step 3: Simplify.

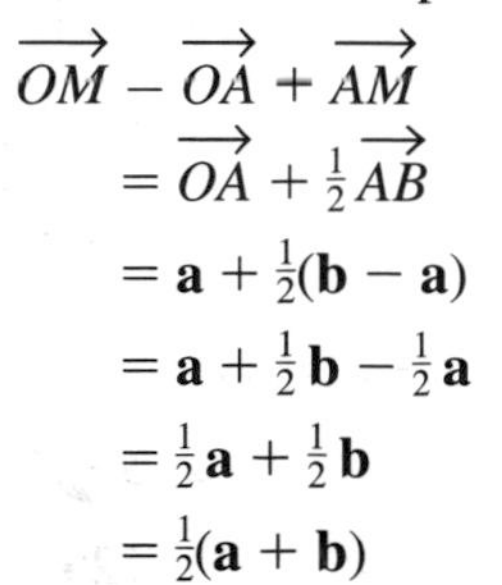

$$\begin{aligned} \overrightarrow{OM} &- \overrightarrow{OA} + \overrightarrow{AM} \\ &= \overrightarrow{OA} + \tfrac{1}{2}\overrightarrow{AB} \\ &= \mathbf{a} + \tfrac{1}{2}(\mathbf{b} - \mathbf{a}) \\ &= \mathbf{a} + \tfrac{1}{2}\mathbf{b} - \tfrac{1}{2}\mathbf{a} \\ &= \tfrac{1}{2}\mathbf{a} + \tfrac{1}{2}\mathbf{b} \\ &= \tfrac{1}{2}(\mathbf{a} + \mathbf{b}) \end{aligned}$$

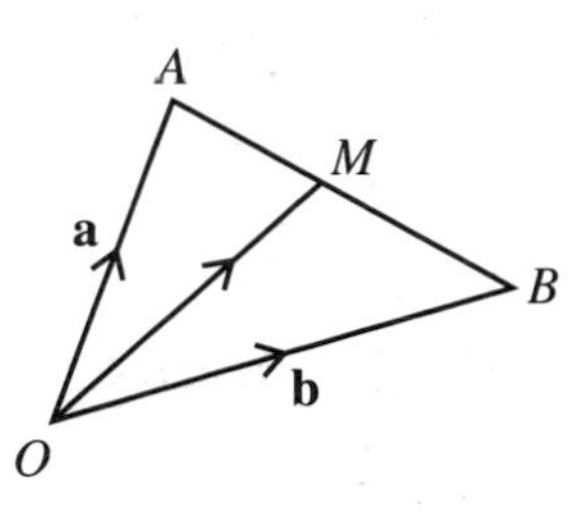

Tip: Use the result $\overrightarrow{AB} = \mathbf{b} - \mathbf{a}$, found earlier.

Note: The position vector of M is $\overrightarrow{OM} = \frac{1}{2}(\mathbf{a} + \mathbf{b})$. This is different from the vector $\overrightarrow{AM}$.

Example 6.6 In the sketch O is a fixed origin. A and B are points with position vectors **a** and **b**. M is the midpoint of OA. Y is the point on AB such that $OY : YB = 3 : 1$. X is the point on AY such that $AX : XY = 4 : 1$.

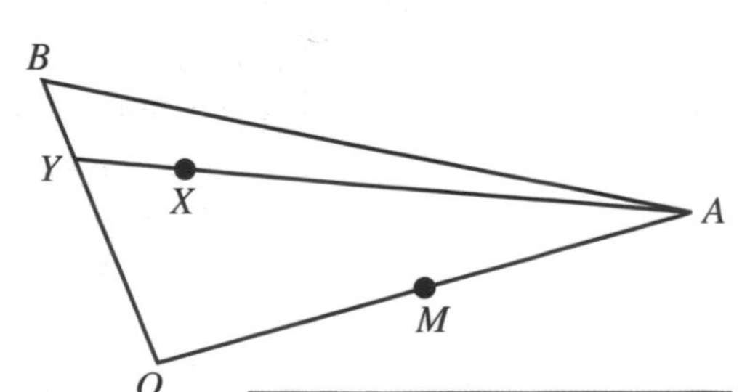

a Find, in terms of **a** and **b**, the position vectors of M and Y.

b Show that $\overrightarrow{MX} = \frac{3}{10}(2\mathbf{b} - \mathbf{a})$.

c Hence show that M, X and B are collinear.

Note: Collinear means lying on the same line.

Step 1: Using the given ratios, find expressions for the position vectors.

a $\overrightarrow{OM} = \frac{1}{2}\overrightarrow{OA} = \frac{1}{2}\mathbf{a}$

$\overrightarrow{OY} = \frac{3}{4}\overrightarrow{OB} = \frac{3}{4}\mathbf{b}$

Tip: If $OY : YB = 3 : 1$, then OY is $\frac{3}{4}$ of the length of OB.

Step 2: Use the triangle law and the given ratios.

b $\overrightarrow{MX} = \overrightarrow{MA} + \overrightarrow{AX}$

$= \frac{1}{2}\overrightarrow{OA} + \frac{4}{5}\overrightarrow{AY}$

Tip: If $AX : XY = 4 : 1$, then AX is $\frac{4}{5}$ of the length of AY.

Step 3: Substitute your answers from part **a**.

$= \frac{1}{2}\mathbf{a} + \frac{4}{5}(\overrightarrow{AO} + \overrightarrow{OY})$

$= \frac{1}{2}\mathbf{a} + \frac{4}{5}(-\mathbf{a} + \frac{3}{4}\mathbf{b})$

Recall: $\overrightarrow{AO} = -\overrightarrow{OA}$ (Section 6.1).

Step 4: Collect terms and simplify.

$= \frac{1}{2}\mathbf{a} - \frac{4}{5}\mathbf{a} + \frac{3}{5}\mathbf{b} = -\frac{3}{10}\mathbf{a} + \frac{3}{5}\mathbf{b}$

$= \frac{3}{10}(2\mathbf{b} - \mathbf{a})$

Tip: Be careful simplifying the fractions. Use the given answer to guide you.

Step 5: Use the triangle law to find $\overrightarrow{MB}$.

c $\overrightarrow{MB} = \overrightarrow{MO} + \overrightarrow{OB}$

$= -\frac{1}{2}\mathbf{a} + \mathbf{b} = \frac{1}{2}(2\mathbf{b} - \mathbf{a}) = \frac{5}{10}(2\mathbf{b} - \mathbf{a})$

Recall: Two vectors are parallel if one is a scalar multiple of the other.

Step 6: Write $\overrightarrow{MX}$ in terms of $\overrightarrow{MB}$ and state your conclusion.

Therefore $\overrightarrow{MX} = \frac{3}{5}\overrightarrow{MB}$

Hence MX and MB are parallel but, since they both have the point M in common, M, X and B are collinear.

The distance between two points

The distance, d, between two points $A(x_1, y_1, z_1)$ and $B(x_2, y_2, z_2)$ is given by

$$d^2 = (x_2 - x_1)^2 + (y_2 - y_1)^2 + (z_2 - z_1)^2$$

Note: This is the three-dimensional version of the formula $d^2 = (x_2 - x_1)^2 + (y_2 - y_1)^2$.

Tip: If you prefer you could work out $\overrightarrow{AB}$ and find its magnitude.

Example 6.7 Given that the length of the line joining $A(2, 4, 1)$ to $B(4, -2, k)$ is 7, find the two possible values of k.

Step 1: Use the formula and simplify.

$AB^2 = (4 - 2)^2 + (-2 - 4)^2 + (k - 1)^2$

$= 2^2 + (-6)^2 + (k^2 - 2k + 1)$

$= 4 + 36 + k^2 - 2k + 1$

$= k^2 - 2k + 41$

Tip: Be careful not to muddle up the signs in this formula.

Step 2: Substitute the given value and simplify.

Since $AB = 7$,

$k^2 - 2k + 41 = 49$

$k^2 - 2k - 8 = 0$

Tip: Don't forget to square the length.

Step 3: Factorise and solve.

$(k - 4)(k + 2) = 0$

$k = 4$ or -2

Tip: If the quadratic expression doesn't factorise, use the quadratic formula.

To check:

The position vectors of A and the two possible points B_1 and B_2 are

$$\overrightarrow{OA} = \begin{pmatrix} 2 \\ 4 \\ 1 \end{pmatrix}, \quad \overrightarrow{OB_1} = \begin{pmatrix} 4 \\ -2 \\ 4 \end{pmatrix} \quad \text{and} \quad \overrightarrow{OB_2} = \begin{pmatrix} 4 \\ -2 \\ -2 \end{pmatrix}$$

so $\overrightarrow{AB_1} = \begin{pmatrix} 2 \\ -6 \\ 3 \end{pmatrix}$ and $\overrightarrow{AB_2} = \begin{pmatrix} 2 \\ -6 \\ -3 \end{pmatrix}$

$$|\overrightarrow{AB_1}| = \sqrt{2^2 + (-6)^2 + 3^2} = \sqrt{49} = 7$$

$$|\overrightarrow{AB_2}| = \sqrt{2^2 + (-6)^2 + (-3)^2} = \sqrt{49} = 7$$

Note:
This check demonstrates the alternative equivalent method of finding the modulus of the vector joining the two points, instead of using the formula.

SKILLS CHECK **6B: Position vectors and distance**

1 Find the vector $\overrightarrow{AB}$ given that the position vectors of A and B relative to O are:

a $\overrightarrow{OA} = 6\mathbf{i} + 2\mathbf{j} - \mathbf{k}$ and $\overrightarrow{OB} = 2\mathbf{i} - 3\mathbf{j} + \mathbf{k}$

b $\mathbf{a} = \mathbf{i} + \mathbf{j} - 4\mathbf{k}$ and $\mathbf{b} = 3\mathbf{i} - 3\mathbf{j} + 2\mathbf{k}$

c $\mathbf{a} = \begin{pmatrix} -3 \\ -1 \\ 2 \end{pmatrix}$ and $\mathbf{b} = \begin{pmatrix} 0 \\ 3 \\ -1 \end{pmatrix}$

2 The points P, Q, and R have position vectors $2\mathbf{a} + \mathbf{b}$, $\mathbf{a} + 3\mathbf{b}$ and $-2\mathbf{a} + k\mathbf{b}$ respectively.

a Find the value of k for which P, Q and R are collinear.

b State the ratio $PQ : QR$.

3 In the diagram A and B have position vectors $\mathbf{a}$ and $\mathbf{b}$, relative to the origin, O. $OA = \frac{3}{4}OC$ and M is the midpoint of OB. The point P lies on CM such that $CP : PM = 2 : 3$.

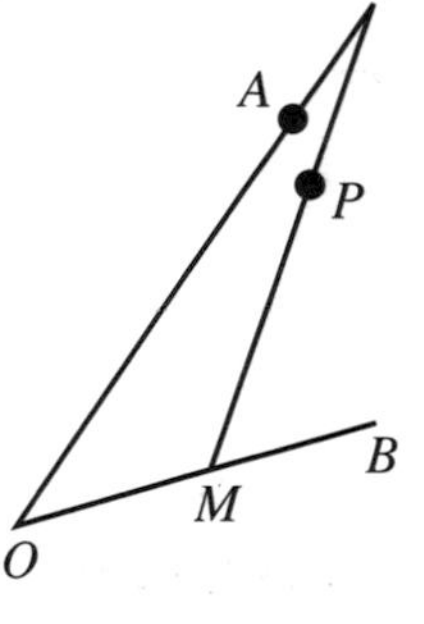

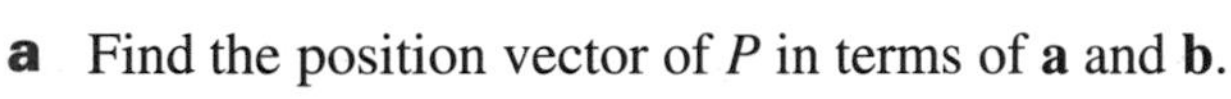

a Find the position vector of P in terms of $\mathbf{a}$ and $\mathbf{b}$.

b Show that A, P and B are collinear.

c Find the ratio $AP : PB$.

4 The position vectors of two points A and B are $\mathbf{a} = 2\mathbf{i} - 4\mathbf{j} + 6\mathbf{k}$ and $\mathbf{b} = 7\mathbf{i} + 11\mathbf{j} - 4\mathbf{k}$ respectively. The point P lies on AB and has position vector $2t\mathbf{i} + t\mathbf{j} + t\mathbf{k}$. Find the ratio $AP : PB$.

5 Find the length of the line joining each of the following pairs of points, giving your answer to one decimal place.

a $A(4, -3, 7)$, $B(-2, -4, 3)$

b $A(-6, 6, 0)$, $B(2, 2, -2)$

c $A(1, 4, -6)$, $B(0, -9, 5)$

6 The points $A(4, 2, 3)$, $B(3, 3, -1)$, $C(6, 0, -1)$ and D form a parallelogram.

a Show that $|\overrightarrow{AB}| = |\overrightarrow{BC}|$.

b Find the position vector of the point D.

7 Given that the distance between the points $A(3, -2, 1)$ and $B(2, -4, p)$ is 3, find both possible values of p.

8 In the diagram $\overrightarrow{OA} = \mathbf{a}$ and $\overrightarrow{OB} = \mathbf{b}$.
A lies on the line OC such that $OA : AC = 1 : 2$.
D lies on the line AB such that $AD : DB = 1 : 2$.
O, D and E are collinear with $OD : DE = 5 : 4$.

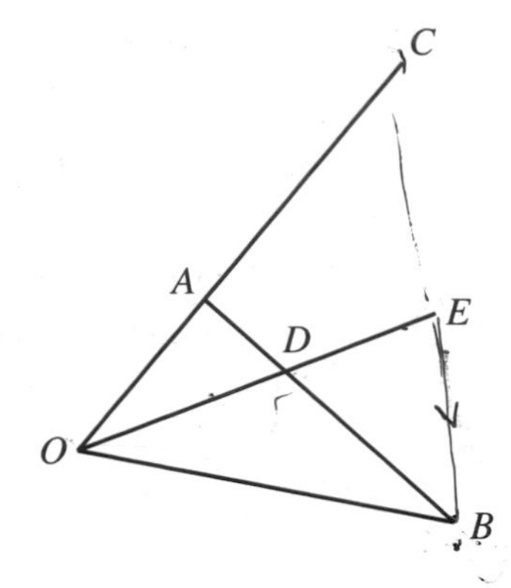

a Find, in terms of **a** and **b**,

i $\overrightarrow{OC}$ **ii** $\overrightarrow{OD}$ **iii** $\overrightarrow{OE}$ **iv** $\overrightarrow{BE}$

b Show that B, E and C are collinear.

c Find the ratio $BE : EC$.

SKILLS CHECK **6B EXTRA is on the CD**

6.3 Vector equations of lines

Vector equations of lines.

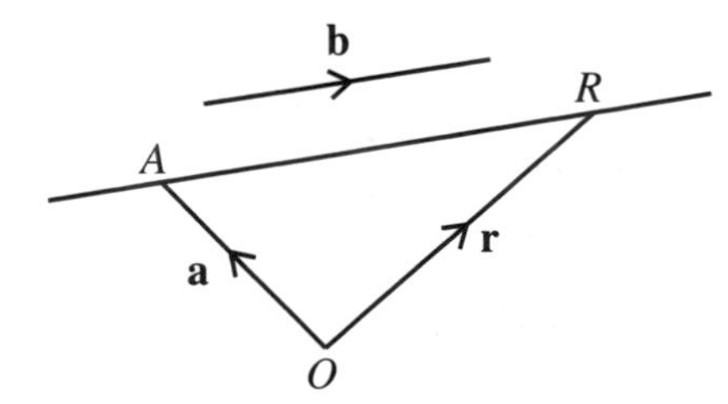

The diagram shows a line which passes through a point A, with position vector **a**, and which is parallel to a vector **b**.

Let **r** be the position vector of a point R on the line.
Since the line is parallel to **b**, then

$$\overrightarrow{AR} = t\mathbf{b}, \quad \text{where } t \text{ is a scalar}$$

$$\mathbf{r} = \overrightarrow{OR} = \overrightarrow{OA} + \overrightarrow{AR}$$

$$\mathbf{r} = \mathbf{a} + t\mathbf{b}$$

- A vector equation of a straight line passing through a point A and parallel to a vector **b** is

$$\mathbf{r} = \mathbf{a} + t\mathbf{b}$$

Different values of the scalar t will give the position vectors of different points that lie on the line.

Note:
The vector **b** is called the **direction vector** of the line.

Note:
There are an infinite number of equations for a particular line: any point on the line could be used in place of A and any multiple of **b** could be used as the direction vector.

Example 6.8 **a** Find a vector equation of the straight line which passes through the point $A(3, -1, 2)$ and is parallel to the vector $4\mathbf{i} - \mathbf{k}$.

b Show that the point P, with position vector $11\mathbf{i} - \mathbf{j}$, lies on the line.

Step 1: State **a**, the position vector of A, and **b**, the direction vector of the line.

a $\mathbf{a} = \begin{pmatrix} 3 \\ -1 \\ 2 \end{pmatrix}, \mathbf{b} = \begin{pmatrix} 4 \\ 0 \\ -1 \end{pmatrix}$

Step 2: Substitute the position vector of A and the direction vector into the vector equation of a line.

An equation of the line is

$$\mathbf{r} = \begin{pmatrix} 3 \\ -1 \\ 2 \end{pmatrix} + t\begin{pmatrix} 4 \\ 0 \\ -1 \end{pmatrix}$$

Recall:
A point with coordinates (x, y, z) has position vector $\begin{pmatrix} x \\ y \\ z \end{pmatrix}$.

This can also be written as

$$\mathbf{r} = 3\mathbf{i} - \mathbf{j} + 2\mathbf{k} + t(4\mathbf{i} - \mathbf{k})$$

$$\text{or } \mathbf{r} = (3 + 4t)\mathbf{i} - \mathbf{j} + (2 - t)\mathbf{k}$$

$$\text{or } \mathbf{r} = \begin{pmatrix} 3+4t \\ -1 \\ 2-t \end{pmatrix}$$

Tip:
Don't forget the zero value for the coefficient of **j**.

Tip:
The last two formats are useful if you need to equate coefficients of **i**, **j** and **k**, as in part **b** of this question.

Step 1: Equate coefficients of **i**, **j** and **k** and attempt to solve for t.

b If P lies on the line then there is a value of t which satisfies

$$\begin{pmatrix} 3+4t \\ -1 \\ 2-t \end{pmatrix} = \begin{pmatrix} 11 \\ -1 \\ 0 \end{pmatrix}$$

i.e. $3 + 4t = 11$

$-1 = -1$

$2 - t = 0$

Since $t = 2$ satisfies all three equations, the point P lies on the line.

The diagram shows a line which passes through two points C and D, with position vectors **c** and **d** respectively.

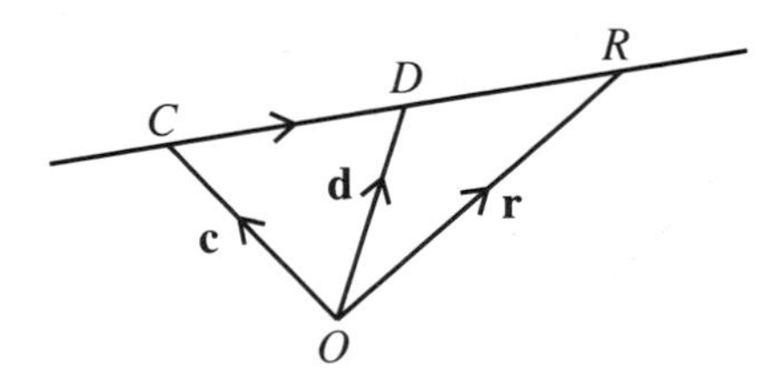

Let **r** be the position vector of a point R on the line.
Since the line passes through C and D, the direction vector of the line is $\overrightarrow{CD} = \mathbf{d} - \mathbf{c}$.

Therefore

$$\mathbf{r} = \overrightarrow{OR} = \overrightarrow{OC} + \overrightarrow{CR}$$

$$\mathbf{r} = \mathbf{c} + t(\mathbf{d} - \mathbf{c})$$

Note:
The vector $\overrightarrow{CR}$ is a scalar multiple of the vector $\overrightarrow{CD}$.

- A vector equation of a straight line passing through the points C and D is

$$\mathbf{r} = \mathbf{c} + t(\mathbf{d} - \mathbf{c})$$

Note:
$\mathbf{r} = \overrightarrow{OD} + \overrightarrow{DR} = \mathbf{d} + t(\mathbf{d} - \mathbf{c})$ would also be an equation of the line.

Example 6.9 **a** Find a vector equation of the straight line which passes through the points $A(0, -3, -3)$ and $B(2, 3, -1)$.

b Given that the point $P(p, 0, q)$ lies on the line, find p and q.

Step 1: Write down the position vectors of the given points.
Step 2: Find a direction vector.

a The position vector of A is $\mathbf{a} = -3\mathbf{j} - 3\mathbf{k}$.
The position vector of B is $\mathbf{b} = 2\mathbf{i} + 3\mathbf{j} - \mathbf{k}$.

$$\overrightarrow{AB} = \mathbf{b} - \mathbf{a}$$

$$= 2\mathbf{i} + 3\mathbf{j} - \mathbf{k} - (-3\mathbf{j} - 3\mathbf{k})$$

$$= 2\mathbf{i} + 6\mathbf{j} + 2\mathbf{k}$$

Recall:
If $\overrightarrow{OA} = \mathbf{a}$ and $\overrightarrow{OB} = \mathbf{b}$, then $\overrightarrow{AB} = \mathbf{b} - \mathbf{a}$ (Section 6.2)

Step 3: Use an appropriate formula for the vector equation of a line.

So an equation of the line is

$$\mathbf{r} = -3\mathbf{j} - 3\mathbf{k} + t(2\mathbf{i} + 6\mathbf{j} + 2\mathbf{k})$$

This can also be written as

$$\mathbf{r} = 2t\mathbf{i} + (-3 + 6t)\mathbf{j} + (-3 + 2t)\mathbf{k}$$

$$\text{or } \mathbf{r} = \begin{pmatrix} 0 \\ -3 \\ -3 \end{pmatrix} + t\begin{pmatrix} 2 \\ 6 \\ 2 \end{pmatrix}$$

$$\text{or } \mathbf{r} = \begin{pmatrix} 2t \\ -3 + 6t \\ -3 + 2t \end{pmatrix}$$

Tip: Use either of the given points to form the equation.

Step 1: Write down the position vector of P.

b $\overrightarrow{OP} = \begin{pmatrix} p \\ 0 \\ q \end{pmatrix}$

If the point P lies on the line then,

Step 2: Set up 3 simultaneous equations by equating coefficients.

$$\begin{pmatrix} 2t \\ -3 + 6t \\ -3 + 2t \end{pmatrix} = \begin{pmatrix} p \\ 0 \\ q \end{pmatrix}$$

Note: The final format from part **a** is easiest to use here.

$p = 2t$ ①

$0 = -3 + 6t$ ②

$q = -3 + 2t$ ③

Tip: Start with the equation that has only one unknown.

Step 3: Solve an equation to find t.

From ②,

$$0 = -3 + 6t$$
$$t = \tfrac{1}{2}$$

Step 4: Substitute t back into the other equations to find p and q.

From ①,

$$p = 2 \times \tfrac{1}{2} = 1$$

From ③,

$$q = -3 + 2 \times \tfrac{1}{2} = -2$$

$p = 1$ and $q = -2$.

Note: So the coordinates of P are $(1, 0, -2)$.

Intersection of lines

In two dimensions two distinct lines are either parallel or they intersect.
In three dimensions a pair of distinct lines may be parallel or intersecting, but they may be neither: such lines are called **skew**.

Example 6.10 Show that the lines with vector equations

$$\mathbf{r} = \mathbf{i} + 4\mathbf{k} + s(-\mathbf{i} + 3\mathbf{j} + 2\mathbf{k})$$

and $$\mathbf{r} = -4\mathbf{j} + 3\mathbf{k} + t(2\mathbf{i} + \mathbf{j} - \mathbf{k})$$

intersect, and find the coordinates of the point of intersection.

Note: Since there are two lines in this question a different parameter is used in each equation.

Step 1: Rewrite the equations in an appropriate format.

$$\mathbf{r} = \begin{pmatrix} 1 - s \\ 3s \\ 4 + 2s \end{pmatrix} \quad \text{and} \quad \mathbf{r} = \begin{pmatrix} 2t \\ -4 + t \\ 3 - t \end{pmatrix}$$

Step 2: Set up three simultaneous equations by equating coefficients.

Equating,

$$\begin{pmatrix} 1-s \\ 3s \\ 4+2s \end{pmatrix} = \begin{pmatrix} 2t \\ -4+t \\ 3-t \end{pmatrix}$$

$$1 - s = 2t \quad ①$$
$$3s = -4 + t \quad ②$$
$$4 + 2s = 3 - t \quad ③$$

Tip: A column vector format makes it easier to set up the equations.

Tip: Solve two of the equations to find the parameters, then check in the third. As an extra check, substitute s and t back into all three equations.

Step 3: Solve two of the equations to find s and t.

Adding ② and ③,

$$4 + 5s = -1$$
$$5s = -5$$
$$s = -1$$

Substituting in ②,

$$3(-1) = -4 + t$$
$$t = 1$$

Step 4: Check that the other equation is satisfied and state a conclusion.

Checking in ①,

$$\text{LHS} = 1 - (-1) = 2$$
$$\text{RHS} = 2(1) = 2$$

Since all three equations are satisfied, the lines intersect when $s = -1$ and $t = 1$.

Tip: You must show that you have checked in the third equation or you will lose marks.

Note: If the lines are skew, the third equation will *not* be satisfied.

Step 5: Substitute one of the parameter values back into the corresponding vector equation.

At the point of intersection $s = -1$ and the position vector of the point of intersection is $\mathbf{r} = \begin{pmatrix} 2 \\ -3 \\ 2 \end{pmatrix}$.

So the coordinates of the point of intersection are $(2, -3, 2)$.

Tip: You could substitute $t = 1$ into the second line equation instead.

Tip: Remember to write your answer as coordinates as requested.

SKILLS CHECK 6C: Vector equations of lines

1 Find a vector equation of the straight line which passes through the point A, with position vector $\mathbf{a}$, and is parallel to the vector $\mathbf{b}$.

a $\mathbf{a} = \begin{pmatrix} 2 \\ 0 \\ -1 \end{pmatrix}$, $\mathbf{b} = \begin{pmatrix} 7 \\ -2 \\ 6 \end{pmatrix}$ **b** $\mathbf{a} = 4\mathbf{i} - \mathbf{j} + 3\mathbf{k}$, $\mathbf{b} = 6\mathbf{j} - \mathbf{k}$

c $\mathbf{a} = -\mathbf{i} + 2\mathbf{j} + \mathbf{k}$, $\mathbf{b} = -\mathbf{i} - \mathbf{j}$

2 Find a vector equation of the straight line which passes through each of the following pairs of points.

a $A(2, -1, 5)$, $B(-3, 0, 1)$ **b** $A(0, 2, 1)$, $B(3, 3, -1)$ **c** $A(1, 4, -2)$, $B(-3, 1, 4)$

3 Show that the point $P(2, -1, 7)$ lies on the line $\mathbf{r} = 4\mathbf{i} - 2\mathbf{j} + 3\mathbf{k} + t(2\mathbf{i} - \mathbf{j} - 4\mathbf{k})$.

4 Given that the point $(a, b, 0)$ lies on the line $\mathbf{r} = 6\mathbf{i} - 4\mathbf{j} + 2\mathbf{k} + \lambda(\mathbf{i} - \mathbf{j} + \mathbf{k})$, find a and b.

5 Prove that the following pairs of lines are skew.

a $\mathbf{r} = 4\mathbf{i} + 9\mathbf{k} + s(\mathbf{i} + 2\mathbf{j} + 5\mathbf{k})$ and $\mathbf{r} = -7\mathbf{i} + 3\mathbf{j} - \mathbf{k} + t(2\mathbf{i} - \mathbf{j} + 3\mathbf{k})$

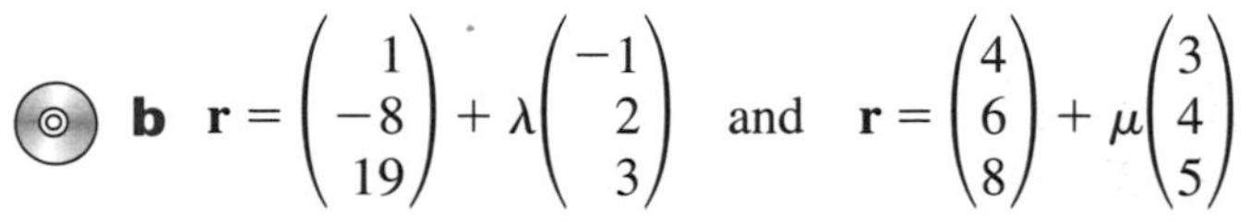

b $\mathbf{r} = \begin{pmatrix} 1 \\ -8 \\ 19 \end{pmatrix} + \lambda\begin{pmatrix} -1 \\ 2 \\ 3 \end{pmatrix}$ and $\mathbf{r} = \begin{pmatrix} 4 \\ 6 \\ 8 \end{pmatrix} + \mu\begin{pmatrix} 3 \\ 4 \\ 5 \end{pmatrix}$

6 Show that the following pairs of lines intersect and find the coordinates of the point of intersection.

a $\mathbf{r} = 2\mathbf{i} + \mathbf{j} - \mathbf{k} + s(\mathbf{i} - \mathbf{j} + 2\mathbf{k})$ and $\mathbf{r} = \mathbf{i} + 2\mathbf{j} - 5\mathbf{k} + t(-\mathbf{i} + \mathbf{j} - \mathbf{k})$

b $\mathbf{r} = \begin{pmatrix} 4 \\ -1 \\ 2 \end{pmatrix} + \lambda \begin{pmatrix} 2 \\ 2 \\ -5 \end{pmatrix}$ and $\mathbf{r} = \begin{pmatrix} 3 \\ -5 \\ 6 \end{pmatrix} + \mu \begin{pmatrix} 1 \\ -2 \\ -1 \end{pmatrix}$

7 **a** Show that the points $A(-3, 6, 0)$ and $B(7, -9, -5)$ lie on the line with equation $\mathbf{r} = \mathbf{i} - 2\mathbf{k} + \lambda(2\mathbf{i} - 3\mathbf{j} - \mathbf{k})$.

b Find the length of AB.

8 **a** Find a vector equation of the line l which passes through the points $A(2, 1, -2)$ and $B(-3, 4, 1)$.

b Given that the point $C(p, -11, q)$ lies on l, find p and q.

SKILLS CHECK **6C EXTRA is on the CD**

6.4 Scalar product

The scalar product. Its use for calculating the angle between two lines.

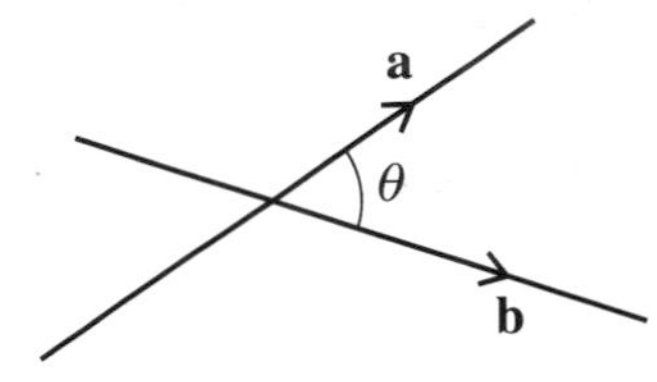

Note:
a.b. is said 'a dot b'. It is sometimes called the dot product.

Note:
It is called the scalar product because it gives a scalar result.

Note:
$\mathbf{a.b} = |\mathbf{a}||\mathbf{b}|\cos\theta = |\mathbf{b}||\mathbf{a}|\cos\theta = \mathbf{b.a}$

The scalar product of two vectors **a** and **b** is written **a.b** and is defined as

$$\mathbf{a.b} = |\mathbf{a}||\mathbf{b}|\cos\theta$$

where θ is the angle between the vectors.

If **a** and **b** are position vectors then $\theta = \angle AOB$.

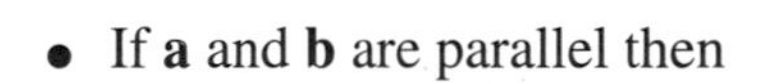

- If **a** and **b** are parallel then

$$\mathbf{a.b} = |\mathbf{a}||\mathbf{b}|\cos 0° = |\mathbf{a}||\mathbf{b}|$$

- If **a** and **b** are perpendicular then

$$\mathbf{a.b} = |\mathbf{a}||\mathbf{b}|\cos 90° = 0$$

- In reverse, if

$$\mathbf{a.b} = 0$$

then two non-zero vectors **a** and **b** are perpendicular.

Recall:
$\cos 0° = 1$, $\cos 90° = 0$ (C2 Section 4.5).

When **a** and **b** are given in component form such that

$$\mathbf{a} = a_1\mathbf{i} + a_2\mathbf{j} + a_3\mathbf{k}$$
$$\mathbf{b} = b_1\mathbf{i} + b_2\mathbf{j} + b_3\mathbf{k}$$

then

$$\mathbf{a.b} = \begin{pmatrix} a_1 \\ a_2 \\ a_3 \end{pmatrix} . \begin{pmatrix} b_1 \\ b_2 \\ b_3 \end{pmatrix} = a_1b_1 + a_2b_2 + a_3b_3$$

Example 6.11 Find the angle between the vectors **a** and **b** when

a $\mathbf{a} = \begin{pmatrix} 8 \\ -1 \\ 4 \end{pmatrix}, \mathbf{b} = \begin{pmatrix} 2 \\ 1 \\ -1 \end{pmatrix}$ **b** $\mathbf{a} = 4\mathbf{i} + 5\mathbf{j} + 3\mathbf{k}, \mathbf{b} = 3\mathbf{i} - 5\mathbf{j} - 4\mathbf{k}$

Step 1: Calculate **a.b**. **a** $\mathbf{a} = \begin{pmatrix} 8 \\ -1 \\ 4 \end{pmatrix}, \quad \mathbf{b} = \begin{pmatrix} 2 \\ 1 \\ -1 \end{pmatrix}$

$$\mathbf{a.b} = \begin{pmatrix} 8 \\ -1 \\ 4 \end{pmatrix} . \begin{pmatrix} 2 \\ 1 \\ -1 \end{pmatrix} = (8 \times 2) + (-1 \times 1) + (4 \times -1) = 11$$

Step 2: Find the modulus of the two vectors.

$$|\mathbf{a}| = \sqrt{8^2 + (-1)^2 + 4^2} = \sqrt{81} = 9$$
$$|\mathbf{b}| = \sqrt{2^2 + 1^2 + (-1)^2} = \sqrt{6}$$

Step 3: Use the definition of the scalar product.

$$\mathbf{a.b} = |\mathbf{a}||\mathbf{b}|\cos\theta$$
$$11 = 9\sqrt{6}\cos\theta$$

Step 4: Rearrange to find $\cos\theta$ and solve.

$$\cos\theta = \frac{11}{9\sqrt{6}}$$
$$\theta = 60.1° \text{ (1 d.p.)}$$

Step 1: Calculate **a.b**. **b** $\mathbf{a} = 4\mathbf{i} + 5\mathbf{j} + 3\mathbf{k}, \quad \mathbf{b} = 3\mathbf{i} - 5\mathbf{j} - 4\mathbf{k}$

$$\mathbf{a.b} = (4 \times 3) + (5 \times -5) + (3 \times -4) = -25$$

Step 2: Find the modulus of the two vectors.

$$|\mathbf{a}| = \sqrt{4^2 + 5^2 + 3^2} = \sqrt{50}$$
$$|\mathbf{b}| = \sqrt{3^2 + (-5)^2 + (-4)^2} = \sqrt{50}$$

Step 3: Use the definition of the scalar product.

$$\mathbf{a.b} = |\mathbf{a}||\mathbf{b}|\cos\theta$$
$$-25 = \sqrt{50}\sqrt{50}\cos\theta$$

Step 4: Rearrange to find $\cos\theta$ and solve.

$$\cos\theta = -\frac{25}{\sqrt{50}\sqrt{50}}$$
$$\theta = 120°$$

Tip: Show some working in case you make a slip.

Recall: Modulus of a vector (Section 6.1).

Tip: Leave your answers in surd form where necessary: using decimals at this stage could lead to an accuracy error later.

Tip: If the question doesn't state the required degree of accuracy leave angles to 1 d.p.

Tip: The scalar product gives a scalar result: your answer should be a number with no **i**, **j** or **k**.

Note: In part **a**, $\cos\theta$ was positive so the angle between the vectors was acute. Here $\cos\theta$ is negative so the angle must be obtuse.

Example 6.12 Given that the vector $\mathbf{a} = 6\mathbf{i} + a\mathbf{j} + 5\mathbf{k}$ is perpendicular to the vector $\mathbf{b} = 2\mathbf{i} + \mathbf{j} - 2\mathbf{k}$, find a.

Step 1: Calculate **a.b**. $\mathbf{a} = 6\mathbf{i} + a\mathbf{j} + 5\mathbf{k}, \quad \mathbf{b} = 2\mathbf{i} + \mathbf{j} - 2\mathbf{k}$

$$\mathbf{a.b} = (6 \times 2) + (a \times 1) + (5 \times -2) = 12 + a - 10 = a + 2$$

Step 2: Use the property of the scalar product of perpendicular vectors. If **a** is perpendicular to **b**,

$$\mathbf{a.b} = 0$$
$$\Rightarrow \quad a + 2 = 0$$

Step 3: Solve for a.

$$a = -2$$

Recall: Perpendicular vectors have a zero scalar product.

Calculating the angle between two lines

Whether two lines intersect or not, the angle between them is given by the angle between their direction vectors.
From the definition of scalar product, the cosine of the angle between two lines is found by

$$\cos\theta = \frac{\mathbf{a.b}}{|\mathbf{a}||\mathbf{b}|}$$

where **a** and **b** are the direction vectors of the lines.

Recall: **b** is the direction vector of the line $\mathbf{r} = \mathbf{a} + t\mathbf{b}$ (Section 6.3).

Note: This is a rearrangement of the definition of scalar product used earlier.

Example 6.13 The vector equations of two straight lines are

$$\mathbf{r} = \mathbf{i} + 2\mathbf{j} - \mathbf{k} + \lambda(3\mathbf{i} - \mathbf{j} + 4\mathbf{k})$$

and $\mathbf{r} = 2\mathbf{i} + 3\mathbf{j} - \mathbf{k} + \mu(-2\mathbf{i} + 3\mathbf{j} + \mathbf{k})$

Find the acute angle between the lines.

Step 1: Calculate **a.b**.

The lines have direction vectors $\mathbf{a} = 3\mathbf{i} - \mathbf{j} + 4\mathbf{k}$ and $\mathbf{b} = -2\mathbf{i} + 3\mathbf{j} + \mathbf{k}$.

$$\mathbf{a.b} = (3 \times -2) + (-1 \times 3) + (4 \times 1) = -6 - 3 + 4 = -5$$

Step 2: Find the modulus of the two vectors.

$$|\mathbf{a}| = \sqrt{3^2 + (-1)^2 + 4^2} = \sqrt{26}$$

$$|\mathbf{b}| = \sqrt{(-2)^2 + 3^2 + 1^2} = \sqrt{14}$$

Step 3: Use the definition of the scalar product.

$$\cos\theta = \frac{\mathbf{a.b}}{|\mathbf{a}||\mathbf{b}|} = -\frac{5}{\sqrt{26}\sqrt{14}}$$

Step 4: Find θ.

$$\theta = 105.19\ldots^\circ$$

Step 5: Calculate the acute angle.

Acute angle $= 180^\circ - 105.19\ldots^\circ = 74.8^\circ$ (1 d.p.)

Tip:
Always calculate **a.b** first. If this is zero the lines are perpendicular and you need go no further.

Note:
As $\cos\theta$ is negative you have calculated the obtuse angle between the lines.

SKILLS CHECK **6D: Scalar product**

1 Find the angle between each of the following pairs of vectors.

a $\mathbf{a} = \begin{pmatrix} 2 \\ 0 \\ 1 \end{pmatrix}, \mathbf{b} = \begin{pmatrix} -3 \\ -5 \\ 2 \end{pmatrix}$

b $\mathbf{a} = -\mathbf{i} + 2\mathbf{j} - \mathbf{k}, \mathbf{b} = -2\mathbf{i} - \mathbf{j} + 3\mathbf{k}$

c $\mathbf{a} = 3\mathbf{i} + 2\mathbf{j} - 4\mathbf{k}, \mathbf{b} = \mathbf{i} - 3\mathbf{j} - 2\mathbf{k}$

2 Given that $\overrightarrow{OA} = 3\mathbf{i} - 7\mathbf{j} + \mathbf{k}$ and $\overrightarrow{OB} = 4\mathbf{i} + \mathbf{j} - 5\mathbf{k}$, show that $\overrightarrow{OA}$ and $\overrightarrow{OB}$ are perpendicular.

3 Use a vector method to find the area of triangle AOB given that O is the origin, A is the point $(1, -3, 2)$ and B is the point $(-4, -1, 3)$. Give your answer to three significant figures.

4 Find the acute angle between the following pairs of lines.

a $\mathbf{r} = \mathbf{i} + 2\mathbf{j} - 2\mathbf{k} + s(2\mathbf{i} - 3\mathbf{j} + 6\mathbf{k})$

$\mathbf{r} = 6\mathbf{i} + 4\mathbf{j} - 3\mathbf{k} + t(\mathbf{i} + 2\mathbf{j} - 2\mathbf{k})$

b $\mathbf{r} = \begin{pmatrix} 7 \\ -8 \\ -2 \end{pmatrix} + \lambda\begin{pmatrix} 3 \\ 2 \\ 6 \end{pmatrix}$ and $\mathbf{r} = \begin{pmatrix} 3 \\ -1 \\ 4 \end{pmatrix} + \mu\begin{pmatrix} 1 \\ 2 \\ 3 \end{pmatrix}$

5 Given that $\mathbf{a} = \begin{pmatrix} 0 \\ t \\ 1 \end{pmatrix}, \mathbf{b} = \begin{pmatrix} 3 \\ 2 \\ 0 \end{pmatrix}$, and the angle between the vectors is 60°, find the exact value of t.

6 Use a vector method to find all the angles of triangle ABC, where A, B and C are the points $(4, -1, 4)$, $(-2, 3, 5)$ and $(1, 0, -6)$, respectively. Give your answers to one decimal place.

7 Given that $\overrightarrow{OP} = \begin{pmatrix} 2 \\ 3 \\ 5 \end{pmatrix}, \overrightarrow{OQ} = \begin{pmatrix} 3 \\ -1 \\ 2 \end{pmatrix}$ and $\overrightarrow{OR} = \begin{pmatrix} 10 \\ -3 \\ 7 \end{pmatrix}$,

a show that $\overrightarrow{QP}$ is perpendicular to $\overrightarrow{QR}$,

b find the area of triangle PQR, leaving your answer in simplified surd form.

8 The lines with equations $\mathbf{r} = \begin{pmatrix} 2 \\ 1 \\ 0 \end{pmatrix} + \lambda\begin{pmatrix} -1 \\ 4 \\ 2 \end{pmatrix}$ and $\mathbf{r} = \begin{pmatrix} 1 \\ 5 \\ 2 \end{pmatrix} + \mu\begin{pmatrix} 2 \\ 3 \\ -5 \end{pmatrix}$ intersect at the point X.

a Show that the lines are perpendicular.

b Find the coordinates of the point X.

9 The vectors $\mathbf{a} = t\mathbf{i} - 5\mathbf{j} + 3\mathbf{k}$ and $\mathbf{b} = 2t\mathbf{i} + t\mathbf{j} - \mathbf{k}$ are perpendicular. Find the possible values of t.

10 The point P is on the line l, with equation $\mathbf{r} = \mathbf{i} - 3\mathbf{j} + \lambda(2\mathbf{i} + \mathbf{j} - 2\mathbf{k})$. Given that $\overrightarrow{OP}$, the position vector of P, is perpendicular to the line l, find the coordinates of P.

SKILLS CHECK **6D EXTRA is on the CD**

Examination practice 6: Vectors

1 The line l_1 has vector equation

$$\mathbf{r} = \begin{pmatrix} 3 \\ 1 \\ 2 \end{pmatrix} + \lambda\begin{pmatrix} 1 \\ -1 \\ 4 \end{pmatrix}$$

and the line l_2 has vector equation

$$\mathbf{r} = \begin{pmatrix} 0 \\ 4 \\ -2 \end{pmatrix} + \mu\begin{pmatrix} 1 \\ -1 \\ 0 \end{pmatrix},$$

where λ and μ are parameters.

The lines l_1 and l_2 intersect at the point B and the acute angle between l_1 and l_2 is θ.

a Find the coordinates of B.

b Find the value of $\cos\theta$, giving your answer as a simplified fraction.

The point A, which lies on l_1, has position vector $\mathbf{a} = 3\mathbf{i} + \mathbf{j} + 2\mathbf{k}$.
The point C, which lies on l_2, has position vector $\mathbf{c} = 5\mathbf{i} - \mathbf{j} - 2\mathbf{k}$.
The point D is such that $ABCD$ is a parallelogram.

c Show that $|\overrightarrow{AB}| = |\overrightarrow{BC}|$.

d Find the position vector of the point D. [Edexcel June 2005]

2 The point A has coordinates $(-3, 0, -7)$ and the point B has coordinates $(-1, 4, 1)$.
The line l has vector equation

$$\mathbf{r} = \mathbf{i} + 2\mathbf{j} - \mathbf{k} + \lambda(2\mathbf{i} + \mathbf{j} + 3\mathbf{k}), \quad \text{where } \lambda \text{ is a parameter.}$$

a Show that the point A lies on the line l.

b Find the distance between the points A and B.

c Find the acute angle between the line l and $\overrightarrow{AB}$, giving your answer to three significant figures.

3 The point A has position vector $\overrightarrow{OA} = -\mathbf{i} - 5\mathbf{j} + 9\mathbf{k}$ and the point B has position vector $\overrightarrow{OB} = 11\mathbf{i} + 13\mathbf{j} - 9\mathbf{k}$. The point P lies on AB and is such that OP is perpendicular to AB.

Find the position vector of P.

4 The equations of the lines l_1 and l_2 are given by

$$l_1: \quad \mathbf{r} = \mathbf{i} + 3\mathbf{j} + 5\mathbf{k} + \lambda(\mathbf{i} + 2\mathbf{j} - \mathbf{k}),$$
$$l_2: \quad \mathbf{r} = -2\mathbf{i} + 3\mathbf{j} - 4\mathbf{k} + \mu(2\mathbf{i} + \mathbf{j} + 4\mathbf{k}),$$

where λ and μ are parameters.

a Show that l_1 and l_2 intersect and find the coordinates of Q, their point of intersection.

b Show that l_1 is perpendicular to l_2.

The point P with x-coordinate 3 lies on the line l_1 and the point R with x-coordinate 4 lies on the line l_2.

c Find, in its simplest form, the exact area of the triangle PQR. [Edexcel Jan 2004]

5 Relative to a fixed origin O, the points A and B have position vectors $(2\mathbf{i} + \mathbf{j} - 3\mathbf{k})$ and $(3\mathbf{i} - 2\mathbf{k})$ respectively.

a Find a vector equation of the line l_1 which passes through A and B.

The line l_2 has equation $\mathbf{r} = (6\mathbf{i} + 4\mathbf{j} - 5\mathbf{k}) + \mu(4\mathbf{i} + 3\mathbf{j} - 2\mathbf{k})$, where μ is a parameter.

b Show that A lies on l_2.

c Find the acute angle between the lines l_1 and l_2, giving your answer to the nearest degree.

6 Two submarines are travelling in straight lines through the ocean. Relative to a fixed origin, the vector equations of the two lines, l_1 and l_2, along which they travel are

$$\mathbf{r} = 3\mathbf{i} + 4\mathbf{j} - 5\mathbf{k} + \lambda(\mathbf{i} - 2\mathbf{j} + 2\mathbf{k})$$
$$\text{and} \quad \mathbf{r} = 9\mathbf{i} + \mathbf{j} - 2\mathbf{k} + \mu(4\mathbf{i} + \mathbf{j} - \mathbf{k}),$$

where λ and μ are scalars.

a Show that the submarines are moving in perpendicular directions.

b Given that l_1 and l_2 intersect at the point A, find the position vector of A.

The point b has position vector $10\mathbf{j} - 11\mathbf{k}$.

c Show that only one of the submarines passes through the point B.

d Given that 1 unit on each coordinate axis represents 100 m, find, in km, the distance AB. [Edexcel June 2001]

7 Referred to a fixed origin O, the points A and B have position vectors $3\mathbf{i} - \mathbf{j} + 2\mathbf{k}$ and $-\mathbf{i} + \mathbf{j} + 9\mathbf{k}$ respectively.

a Show that OA is perpendicular to AB.

b Find in vector form, an equation of the line L_1 which passes through A and B.

The line L_2 has equation $r = (8\mathbf{i} + \mathbf{j} - 6\mathbf{k}) + \mu(\mathbf{i} - 2\mathbf{j} - 2\mathbf{k})$, where μ is a scalar parameter.

c Show that the lines L_1 and L_2 intersect and find the position vector of their point of intersection.

d Calculate, to the nearest tenth of a degree, the acute angle between L_1 and L_2. [Edexcel P3 Specimen]

8 Relative to a fixed origin O, the point A has position vector $\mathbf{i} - 2\mathbf{k}$ and the point B has position vector $-\mathbf{j} - 2\mathbf{k}$.

a Find the exact value of the cosine of angle AOB.

b Find the exact value of the area of triangle AOB.

c Find the coordinates of the midpoint of AB.

 **9** Relative to a fixed origin O, the vector equations of the two lines l_1 and l_2 are

$$l_1: \mathbf{r} = 7\mathbf{i} + t\mathbf{j} - 9\mathbf{k} + \lambda(3\mathbf{i} - 5\mathbf{j} - 4\mathbf{k})$$

and $$l_2: \mathbf{r} = 2\mathbf{j} - 8\mathbf{k} + \mu(4\mathbf{i} + 5\mathbf{j} + 3\mathbf{k})$$

where λ, μ are parameters and t is a constant.

The two lines intersect at the point A.

a Find the value of t.

b Find the position vector of the point A.

c Prove that the angle between l_1 and l_2 is 120°.

Point P lies on l_2. Given that the length of AP is $10\sqrt{2}$,

d find the two possible position vectors for the point P.

10 The point A has position vector $3\mathbf{i} + 5\mathbf{j} - \mathbf{k}$ and the point B has position vector $2\mathbf{i} + 18\mathbf{j} - 14\mathbf{k}$.

a Find a vector equation for the line l_1 which passes through A and B.

The line l_2 has equation $\mathbf{r} = 2\mathbf{i} - 4\mathbf{j} + 6\mathbf{k} + t(\mathbf{i} - 2\mathbf{j} + 3\mathbf{k})$.

b Show that l_2 passes through the origin, O.

c Show that the lines l_1 and l_2 intersect at a point P and find the position vector of P.

d Find the size of angle AOP, giving your answer to one decimal place.

e Find the area of triangle AOP, giving your answer to one decimal place.

11 The vector equations of two straight lines are

$$\mathbf{r} = 2\mathbf{i} - 3\mathbf{j} + \mathbf{k} + \lambda(3\mathbf{i} - \mathbf{j} - 2\mathbf{k})$$

and $$\mathbf{r} = t\mathbf{i} + \mathbf{j} + 8\mathbf{k} + \mu(2\mathbf{i} + 2\mathbf{j} + 3\mathbf{k})$$

Given that the two lines intersect, find

a the coordinates of the point of intersection,

b the value of the constant t,

c the acute angle between the two lines.

 **12** Relative to a fixed origin O, the point A has position vector $4\mathbf{j} - 2\mathbf{k}$ and the point B has position vector $3\mathbf{i} - 2\mathbf{j} + 7\mathbf{k}$.

a Find a vector equation of the line l which passes through A and B.

The point P lies on the line l and is such that OP is perpendicular to l.

b Find the position vector of P.

Given that the points O, A, B and C form a parallelogram $OABC$,

c state the position vector of C,

d find the area of the parallelogram $OABC$.

Practice exam paper

Answer **all** questions.
Time allowed: 1 hour 30 minutes
A calculator is **allowed** in this paper.

1 Find $\int x \sin 2x \, dx$. *(5 marks)*

2 The table below is a table of values of $y = \sqrt{(2 + e^{0.5x})}$ to three decimal places.

x	0	0.5	1	1.5	2
y	1.732	1.812	p	2.029	q

a Find the value of p and the value of q. *(2 marks)*

b Use the trapezium rule and all the values of y in the completed table to estimate the value of

$$\int_0^2 \sqrt{(2 + e^{0.5x})} \, dx.$$ *(4 marks)*

3 The population p of bacteria in a culture at time t minutes ($t \geqslant 0$) is modelled using the equation $p = p_0(1.1)^t$. Given that p satisfies the differential equation $\frac{dp}{dt} = kp$,

a show that $k = \ln 1.1$. *(4 marks)*

Given that $p_0 = 3 \times 10^6$,

b calculate the rate of increase of the population when $t = 30$. Give your answer to three significant figures. *(3 marks)*

4 **a** Using the substitution $u = 8x^3 - 1$, find $\int x^2(8x^3 - 1)^4 \, dx$. *(4 marks)*

Figure 1

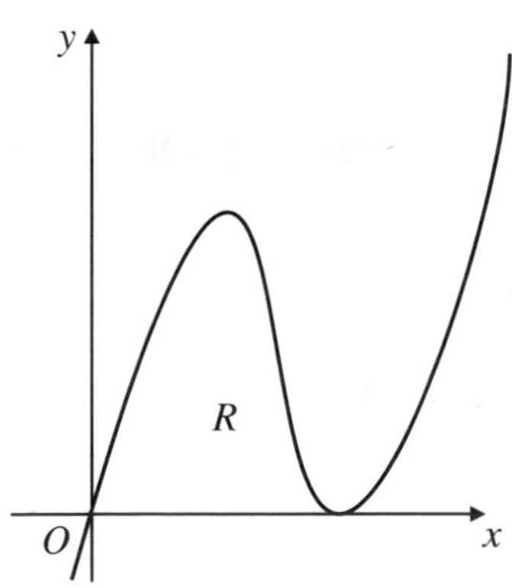

Figure 1 shows part of the curve with equation $y = 3x(8x^3 - 1)^2$. The region R is the finite region bounded by the curve and the x-axis. The region R is rotated through 2π radians about the x-axis.

b Find the exact value of the volume generated. *(5 marks)*

5 **a** Use the binomial theorem to expand $(2 - x)^{-2}$, where $-\frac{1}{2} < x < \frac{1}{2}$, in ascending powers of x, up to and including the term in x^3, simplifying each term. *(5 marks)*

b Expand $\frac{4 + x}{(2 - x)^2}$, where $-\frac{1}{2} < x < \frac{1}{2}$, in ascending powers of x, up to and including the term in x^2, simplifying each term. *(4 marks)*

6 Relative to a fixed origin O, the point A has position vector $-2\mathbf{i} + 3\mathbf{j} - 4\mathbf{k}$ and the point B has position vector $4\mathbf{i} - 3\mathbf{j} - \mathbf{k}$. The points A and B lie on the straight line l. Find

a the vector $\overrightarrow{AB}$, *(2 marks)*

b a vector equation of l. *(2 marks)*

The straight line m has vector equation $\mathbf{r} = \mathbf{i} - 3\mathbf{j} + \mathbf{k} + \mu(-\mathbf{i} + 4\mathbf{j} - 4\mathbf{k})$.

c Show that l and m intersect. *(5 marks)*

d Find the acute angle between l and m. *(4 marks)*

7 **Figure 2**

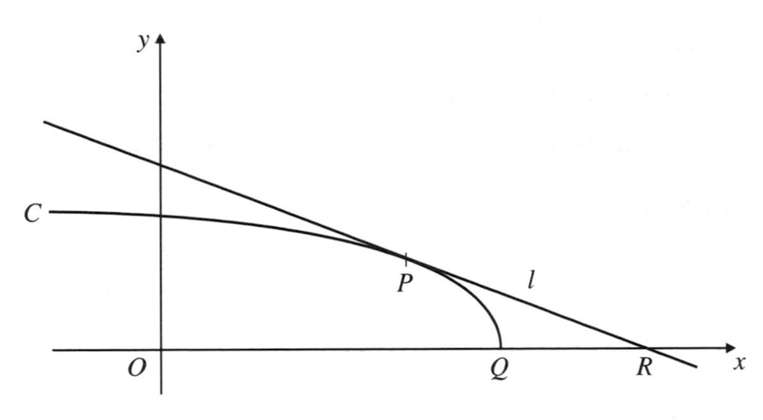

Figure 2 shows the curve C with parametric equations

$$x = 3\cos 2t + 1,\ y = 2\sin t,\ 0 < t \leqslant \frac{\pi}{2}.$$

The point P is where $t = \frac{\pi}{4}$. The line l is the tangent to C at P.

a Show that $\frac{\mathrm{d}y}{\mathrm{d}x} = -\frac{1}{6\sin t}$. *(5 marks)*

b Find an equation for l. *(4 marks)*

The point Q is the point at which C meets the x-axis and the point R is the point at which l cuts the x-axis.

c Calculate the distance QR. *(4 marks)*

8 $$\mathrm{f}(x) = \frac{2x^2}{(2x+1)(x+1)},\quad x \in \mathbb{R}, x \neq -\tfrac{1}{2}, x \neq -1.$$

Given that $\mathrm{f}(x) = A + \frac{B}{2x+1} + \frac{C}{x+1}$,

a find the values of the constants A, B and C. *(4 marks)*

Given that $(2x+1)(x+1)\frac{\mathrm{d}y}{\mathrm{d}x} = 2yx^2$ and that $y = \mathrm{e}$ at $x = 0$,

b find the value of y at $x = 1$. *(9 marks)*

Answers

SKILLS CHECK 1A (page 5)

1 **a** $\frac{1}{x-3} - \frac{1}{x+1}$ **b** $\frac{1}{2(x-3)} - \frac{1}{2(3x-5)}$
c $\frac{3}{x+2} - \frac{2}{2x-3}$

2 **a** $\frac{1}{x-1} - \frac{1}{x} - \frac{1}{x^2}$ **b** $\frac{11}{x+2} - \frac{9}{x+1} + \frac{5}{(x+1)^2}$
c $\frac{2}{2x-1} + \frac{1}{x+2} - \frac{3}{(x+2)^2}$

3 **a** $1 - \frac{4}{x-4} + \frac{12}{x-6}$ **b** $x + \frac{1}{2(x-1)} + \frac{1}{2(x+1)}$
c $x + 2 + \frac{8}{x-3} + \frac{5}{x+7}$

4 **a** $\frac{5}{x-3} - \frac{4}{x+3}$ **b** $5 + \frac{2}{x+1} - \frac{3}{2-x}$
c $\frac{3}{4(1-x)} + \frac{3}{4(1+x)} - \frac{3}{2(1+x)^2}$

5 **a** $\frac{2}{1-2x} + \frac{1}{x+1}$ **b** $(-\frac{1}{4}, \frac{8}{3})$

6 **a** $\frac{2}{1-x} - \frac{3}{1+x}$ **b** $\frac{2}{(1-x)^2} + \frac{3}{(1+x)^2}$
c -2

7 **a** $-\frac{1}{x-5} - \frac{2}{2x+3}$

Exam practice 1 (page 6)

1 $\frac{2}{x} - \frac{3}{x+1} + \frac{1}{x-1}$

2 $\frac{2}{x+2} - \frac{1}{(x-1)^2}$

3 $-2 + x + \frac{4}{x+1} + \frac{2}{x+4}$; $A = -2, B = 1, C = 4, D = 2$

4 **a** $\frac{3}{2} + \frac{1}{x-2} - \frac{1}{2(2x-3)}$ **b** $\frac{5}{3}$ and 1

5 **a** $\frac{1}{x+1} + \frac{1}{x-1} - \frac{2}{x}$ **b** $-\frac{11}{18}$

6 **a** $\frac{2}{2x-1} - \frac{1}{x+3}$ **b** $\frac{1}{(x+3)^2} - \frac{4}{(2x-1)^2}$
c $\frac{16}{(2x-1)^3} - \frac{2}{(x+3)^3}$

7 **a** $\frac{1}{x+2} + \frac{2}{x-1}$

SKILLS CHECK 2A (page 11)

1 **a** $(0, 2)$ **b** $(0, -4), (0, 4)$
2 **a** $(6, 0)$ **b** $(-1, 0), (5, 0)$
3 $a = 4$
4 **a** $y = \frac{1}{x-3}$ **b** $y = \frac{\sqrt{1-x^2}}{x}$ **c** $y = 3\sqrt{x^2-1}$
5 $(\frac{3}{5}, \frac{2}{5}), (\frac{3}{2}, -\frac{1}{2})$
6 **a** $\left(\frac{x-2}{3}\right)^2 + \left(\frac{y+4}{3}\right)^2 = 1$ **b** Radius 3, centre $(2, -4)$
7 9
8 $18\frac{1}{6}$

Exam practice 2 (page 12)

1 $c = 3$
2 **a** $(0, 1), (-\frac{1}{\sqrt{2}}, 0), (\frac{1}{\sqrt{2}}, 0)$ **b** $y = 1 - 2x^2$
3 **a** $y = (x+1)^3 + (x+1)^2$ **b** $(-1, 0), (-2, 0)$
4 $k = 3$
5 $(-\frac{\sqrt{3}}{2} - 1, -\frac{\sqrt{3}}{2} + 1), (\frac{\sqrt{3}}{2} - 1, \frac{\sqrt{3}}{2} + 1)$
6 **a** $\left(\frac{x+1}{4}\right)^2 + \left(\frac{y-5}{4}\right)^2 = 1$ **b** Radius 4, centre $(-1, 5)$
7 $y + 2x = 7$
8 **a** $t_1 = -2, t_2 = 2$ **b** $17\frac{1}{15}$

SKILLS CHECK 3A (page 17)

1 **a** $1 + 2x + 3x^2 + 4x^3, |x| < 1$ **b** $1 - 3x - 9x^2 - 45x^3, |x| < \frac{1}{9}$
c $1 + 6x + 30x^2 + 140x^3, |x| < \frac{1}{4}$
2 **a** $-1 - 6x - 24x^2, |x| < \frac{1}{2}$ **b** $\frac{1}{8} - \frac{3}{64}x + \frac{15}{1024}x^2, |x| < 4$
3 $2 - 3x + 4x^2 - 5x^3$
4 $\frac{15}{8}x^2$
5 $1 + \frac{1}{2}x - \frac{1}{8}x^2 + \frac{1}{16}x^3$, 1.7321
6 **a** $\frac{2}{2+x} - \frac{1}{(1-x)^2}$ **b** $-\frac{5}{2}x - \frac{11}{4}x^2 - \frac{33}{8}x^3$ **c** $|x| < 1$
7 **a** $a = -2, n = -1$ **b** 8 **c** $|x| < \frac{1}{2}$

Exam practice 3 (page 18)

1 **a** $\frac{1}{27} - \frac{2}{27}x + \frac{8}{81}x^2 - \frac{80}{729}x^3$ **b** $|x| < \frac{3}{2}$
2 **a** $A = -1, B = -1$ **b** $-\frac{3}{2} - \frac{5}{4}x - \frac{9}{8}x^2 - \frac{17}{16}x^3$
c $|x| < 1$
3 **a** $k = 2, n = -\frac{1}{2}$ **b** $-\frac{5}{2}$ **c** $|x| < \frac{1}{2}$
4 **a** $1 - 6x + 27x^2 - 108x^3$ **b** $4 - 23x + 102x^2$
5 **a** $A = 1, B = 2$ **b** $3 + 5x + 9x^2$ **c** $|x| < \frac{1}{2}$
6 **a** $p = \frac{1}{12}, q = -\frac{1}{288}$ **b** 2.41 **c** 2.15%
7 **a** $a = 6, n = \frac{5}{2}$ **b** $\frac{135}{2}$
8 **a** $\frac{1}{2}x^2 + \frac{1}{4}x^3$
9 **a** $A = 2, B = 16$ **b** $10 + 10x^2 + 15x^3$

SKILLS CHECK 4A (page 23)

1 **a** $\frac{6x+5}{2y+6}$ **b** $\frac{2-2xy}{3y^2+x^2}$ **c** $\frac{6xy - y\ln y}{3y^3 + x}$
2 $(3, 3), (3, 5)$
3 **a** $\frac{x-2}{3-3y}$ **b** $y + 2x = 9$
4 **a** $-2, \frac{1}{2}$ **b** $90°$
5 **a** $\frac{1}{t}$ **b** $\frac{1}{2}\operatorname{cosec} t$ **c** $\frac{\ln 4t + 1}{2t - 4}$
6 **a** $\frac{2}{t}$ **b** $2y + x = 18$ **c** $p = 18, q = 9$
8 $a = 4$
9 $(1, 1)$

SKILLS CHECK 4B (page 29)

1 **a** $5^x \ln 5$ **b** $-(\frac{1}{2})^t \ln 2$
c $xa^x(2 + x\ln a)$ **d** $2(3^{1+2t})\ln 3$
2 **a** $\frac{2^t \ln 2}{\cos t}$
b $y = x\ln 2 + 1$
3 **a** 1.25
b -2.99×10^{15}
c Rate of decrease of number of atoms after 10 days
4 **a** 58 units/s
b Decreasing at a rate of $78\frac{2}{3}$ units/s
5 $0.3\pi\ \text{cm}^2/\text{s}$
6 0.50 cm/s
7 $\frac{dy}{dx} = \frac{x+y}{4}$
8 $\frac{dr}{dt} = k$
9 $\frac{dV}{dt} = \frac{\sqrt[3]{V}}{20}$

Exam practice 4 (page 30)

1 $\frac{dy}{dx} = \frac{2e^{2x} - y}{x - 2e^{2y}}$

2 b $\frac{dy}{dx} = \frac{6t}{2^t \ln 2}$ **c** 1.307

3 a $y - 7 = -(x - 9)$ **b** 58.9

4 b $t = 2$

5 b $y = \frac{b\sqrt{2}}{a}x - b$

6 b $(-2, 1), (2, -1)$

7 a Centre (4, 8), radius 17 **b** $\frac{4 - x}{y - 8}$

c $x = 21$

8 $\frac{dy}{dx} = \frac{5y - 13x}{13y - 5x}$

9 $x - 2y + 2 = 0$

10 a £1975.31 (nearest p) **b** -800.9 (1 d.p.)

c On 1 January 2005 the value of the car is depreciating at a rate of approximately £800 per year.

11 b -0.265 (3 s.f.)

SKILLS CHECK 5A (page 38)

1 a $\frac{1}{3}e^{3x+1} + c$ **b** $e^{-u} + c$ **c** $-\frac{1}{2}e^{-2t} + c$

2 a $\frac{1}{3}\ln|x| + c$ **b** $\frac{1}{5}\ln|1 + 5x| + c$ **c** $-\frac{1}{3}\ln|2 - x| + c$

3 a $2e^{2x} + 2x$ **b** $\frac{1}{2}\ln|e^{2x} + x^2| + c$

4 a $3\sin\frac{1}{3}x + c$ **b** $-\frac{1}{4}\cos 2y + c$ **c** $3\tan x + c$

5 a $\frac{1}{2}$ **b** 81

6 a $\frac{1}{4}$ **b** $\frac{1}{2} - \frac{1}{8}\pi$

7 2

8 $y = \frac{1}{3}e^{3x} - x^2 + \frac{2}{3}$

9 a $\cos^2 A \equiv \frac{1}{2}(1 + \cos 2A)$ **b** $\frac{1}{2}\pi$

10 b $\frac{1}{18}\sin 9x + \frac{1}{6}\sin 3x + c$

11 $a = -7, b = 2$

12 a $\frac{\cos x}{\sin x}$ **c** $-\cot x - x + c$

13 a $e^{3x} - e^{2x} - e^x + 1$ **b** 0.0067 (2 s.f.)

SKILLS CHECK 5B (page 41)

1 a $\ln 2$ **b** $\frac{1}{2}\pi$

2 $\frac{1}{2}\pi(e^2 - 1)$

3 $\frac{127}{7}\pi$

4 $\pi \ln 2$

5 a $A(0, 1), B(1, 2)$ **b** $\frac{7}{15}\pi$

6 a (3, 3)

7 a $\cap$ shaped quadratic curve through $(-3, 0)$, $(0, 9)$ and $(3, 0)$

b 142 (3 s.f.)

8 $\frac{1}{2}\pi^2$

9 a $t = 1, t = 2$ **b** 112 (3 s.f.)

10 $\frac{1}{2}\pi^2$

SKILLS CHECK 5C (page 46)

1 a 227.5

b $12\frac{4}{9}$

2 b 0.75

3 See CD.

4 $\frac{1}{2}e^{x^2+2} + c$

5 $-6\sqrt{9 - x} + c$

6 $-\frac{1}{2}\ln|1 - e^{2x}| + c$

7 $\frac{1}{2}(1 + \ln x)^2 + c$

8 a $\frac{1}{4}\left(\ln|2x + 1| + \frac{1}{2x + 1}\right) + c$

b 0.125 (3 s.f.)

9 a $-\frac{1}{3}\cos^3 x + c$

b $\frac{1}{3}$

10 2

11 $\arcsin x + c$

SKILLS CHECK 5D (page 49)

1 a $(2x + 3)\sin x + 2\cos x + c$ **b** $\frac{1}{2}x\sin(2x + 3) + \frac{1}{4}\cos(2x + 3) + c$

2 a $\frac{1}{3}xe^{3x} - \frac{1}{9}e^{3x} + c$ **b** $\frac{1}{3}xe^{3x} - \frac{4}{9}e^{3x} + c$

3 5.93 (3 s.f.)

4 a $-\frac{5}{2}x\cos 2x + \frac{5}{4}\sin 2x + c$

5 a i 1 **ii** $\frac{1}{4}\pi^2 - 2$ **b** $\frac{1}{4}\pi(\pi^2 - 8)$

6 a $\frac{1}{2}(1 - \cos 2A)$ **b** $\frac{1}{4}\pi^2$

7 a $-\frac{1}{2}xe^{-2x} - \frac{1}{4}e^{-2x} + c$

8 a $x\tan x + \ln|\cos x| + c$

SKILLS CHECK 5E (page 51)

1 a $A = 2, B = -1$ **b** $\frac{2}{3}\ln|3x + 1| - \frac{1}{2}\ln|2x - 5| + c$

2 a $\frac{1}{x + 2} - \frac{4}{2x - 1} + \frac{2}{x + 3}$ **b** $\ln\frac{81}{2}$

3 a $A = 1, B = -\frac{3}{2}, C = \frac{3}{2}$ **b** $x - \frac{3}{2}\ln|x + 3| + \frac{3}{2}\ln|x - 3| + c$

4 a $\frac{1}{x} - \frac{1}{x - 1} + \frac{1}{(x - 1)^2}$ **b** $\ln\frac{3}{4} + \frac{1}{2}$

5 a $A = 1, B = 4, C = 1$ **b** 15.3 (3 s.f.)

SKILLS CHECK 5F (page 54)

1 a $\ln|y + 1| = \frac{1}{2}x^2 + c$ **b** $y = 3e^{\frac{1}{2}x^2} - 1$

2 a $-\frac{1}{y^2} = \frac{1}{3}x^3 + c$

3 $x^2 + y^2 - y - \frac{15}{4} = 0$, centre $(0, \frac{1}{2})$, radius 2

4 a $x = 4t^{\frac{1}{4}} - 3$ **b** $x = 4e^{\frac{1}{4}(t - 1)} - 3$

5 $\tan\theta = \frac{1}{2}x^2 + c$

6 a i $\frac{1}{2}x\sin 2x + \frac{1}{4}\cos 2x + c$ **ii** $\frac{1}{2}x + \frac{1}{4}\sin 2x + c$

b $\frac{1}{2}y + \frac{1}{4}\sin 2y = \frac{1}{2}x\sin 2x + \frac{1}{4}\cos 2x - \frac{1}{8}\pi$

7 a $\frac{1}{15}(9 + x^3)^5 + c$ **b** $y = \ln(9 + x^3)^5$

8 d 56.7 °C (1 d.p.)

SKILLS CHECK 5G (page 57)

1 6.06 (3 s.f.)

2 2.34 (3 s.f.)

3 a 6.34 (3 s.f.)

b Split the area into more strips; the greater the number of strips, the greater the accuracy of the estimate.

4 a $a = 0.848, b = 0.485$ **b** 0.724 (3 s.f.)

c $\frac{1}{2} + \frac{1}{4}\sin 2$ **d** 0.46% (2 s.f.)

5 8.20 (3 s.f.)

Exam practice 5 (page 58)

1 a $e - 1$ **b** $\frac{1}{2}\pi(e^2 - 1)$

2 $y = -\ln(1 + e - e^x)$

3 a $\sin^2 A \equiv \frac{1}{2}(1 - \cos 2A)$ **b** $\frac{1}{2}(x - \frac{1}{2}\sin 2x) + c$

c $\frac{\pi}{12} - \frac{\sqrt{3}}{8}$

4 a $A = \frac{1}{2}, B = 2, C = -1$ **b** $\frac{1}{2}\ln|x| + 2\ln|x - 1| + \frac{1}{x - 1} + c$

5 a $\frac{1}{24}(1 + 2x^2)^6 + c$ **b** $\tan 2y = \frac{1}{12}((1 + 2x^2)^6 + 11)$

6 a $\frac{1}{2}x\sin 2x + \frac{1}{4}\cos 2x + c$

7 a $\frac{\sqrt{3}}{2}\pi$ **c** $6\ln 2 + \frac{1}{3}\pi^2$

9 $\frac{1}{16}$

10 a $y = A(1 + x^2)$ **b** $y = 2(1 + x^2)$

11 a i $\ln|x| - \ln|x+1| + c$ ii $x - \ln(1 + e^x) + c$
b $-x^2 \cos x + 2x \sin x + 2 \cos x + c$
12 a $A = 1, B = \frac{1}{2}, C = -\frac{1}{2}$ b $x + \frac{1}{2}\ln|x-1| - \frac{1}{2}\ln|x+1| = 2t - \frac{1}{2}\ln 3$
13 a 1, 1.46, 1.42, 0 b 1.8 (1 d.p.)
c Underestimate, trapezia are under the curve
14 a $1 + \frac{1}{5}e^x$ b $6x - 5y + 1 = 0$
c 1.24, 1.55, 1.86 d 2.43 (3 s.f.)
15 $\frac{635}{3}\pi$
16 b 1.69, 2.39 c 4.67 (3 s.f.)

SKILLS CHECK 6A (page 66)

1 a Neither b Equal c Parallel
2 $4\mathbf{i} + 5\mathbf{j} - 6\mathbf{k}$
3 a 13 b $\sqrt{14}$ c 5
d $2\sqrt{5}$ e $\sqrt{30}$ f $\sqrt{6}$
4 a $7\mathbf{i} - 5\mathbf{j} - 7\mathbf{k}$ b $3\mathbf{i} + 5\mathbf{j} - 3\mathbf{k}$ c $4\mathbf{i} + \mathbf{j} + 2\mathbf{k}$
d $\sqrt{33}$ e $\sqrt{61}$ f $\sqrt{34}$
5 a $\begin{pmatrix}-7\\-8\\-7\end{pmatrix}$ b $\begin{pmatrix}6\\5\\3\end{pmatrix}$ c $\begin{pmatrix}7\\2\\-2\end{pmatrix}$
d $\sqrt{38}$ e $3\sqrt{17}$ f $3\sqrt{11}$
6 $\frac{\sqrt{3}}{15}\mathbf{i} + \frac{\sqrt{3}}{3}\mathbf{j} - \frac{7\sqrt{3}}{15}\mathbf{k}$
7a i $2\mathbf{i} - 3\mathbf{j}$ ii $3\mathbf{i} + 2\mathbf{j}$ iii
$\mathbf{i} + 5\mathbf{j}$
b i $\sqrt{13}$ ii $\sqrt{13}$ iii $\sqrt{26}$
8 $-1, 3$

SKILLS CHECK 6B (page 69)

1 a $-4\mathbf{i} - 5\mathbf{j} + 2\mathbf{k}$ b $2\mathbf{i} - 4\mathbf{j} + 6\mathbf{k}$ c $\begin{pmatrix}3\\4\\-3\end{pmatrix}$
2 a 9 b 1 : 3
3 a $\frac{4}{5}\mathbf{a} + \frac{1}{5}\mathbf{b}$ c 1 : 4
4 2 : 3
5 a 7.3 b 9.2 c 17.1
6 a $|\overrightarrow{AB}| = |\overrightarrow{BC}| = \sqrt{18}$ b $7\mathbf{i} - \mathbf{j} + 3\mathbf{k}$
7 $-1, 3$
8 a i $3\mathbf{a}$ ii $\frac{2}{3}\mathbf{a} + \frac{1}{3}\mathbf{b}$ iii $\frac{6}{5}\mathbf{a} + \frac{3}{5}\mathbf{b}$ iv $\frac{6}{5}\mathbf{a} - \frac{2}{5}\mathbf{b}$
c 2 : 3

SKILLS CHECK 6C (page 73)

1 a $\mathbf{r} = \begin{pmatrix}2\\0\\-1\end{pmatrix} + \lambda\begin{pmatrix}7\\-2\\6\end{pmatrix}$ b $\mathbf{r} = 4\mathbf{i} - \mathbf{j} + 3\mathbf{k} + \lambda(6\mathbf{j} - \mathbf{k})$
c $\mathbf{r} = -\mathbf{i} + 2\mathbf{j} + \mathbf{k} - \lambda(\mathbf{i} + \mathbf{j})$
2 a $\mathbf{r} = 2\mathbf{i} - \mathbf{j} + 5\mathbf{k} + \lambda(-5\mathbf{i} + \mathbf{j} - 4\mathbf{k})$
b $\mathbf{r} = 2\mathbf{j} + \mathbf{k} + \lambda(3\mathbf{i} + \mathbf{j} - 2\mathbf{k})$
c $\mathbf{r} = \mathbf{i} + 4\mathbf{j} - 2\mathbf{k} + \lambda(-4\mathbf{i} - 3\mathbf{j} + 6\mathbf{k})$
3 $t = -1$
4 $a = 4, b = -2$
6 a $(-1, 4, -7)$ b $(2, -3, 7)$
7 b $5\sqrt{14}$
8 a $\mathbf{r} = 2\mathbf{i} + \mathbf{j} - 2\mathbf{k} + \lambda(-5\mathbf{i} + 3\mathbf{j} + 3\mathbf{k})$
b $p = 22, q = -14$

SKILLS CHECK 6D (page 76)

1 a 106.9° (1 d.p.) b 109.1° (1 d.p.) c 75.6° (1 d.p.)
2 $\mathbf{a.b} = 12 - 7 - 5 = 0$
3 9.21
4 a 40.4° (1 d.p.) b 17.3° (1 d.p.)
5 $\sqrt{\frac{13}{3}}$
6 $\angle BAC = 81.0°, \angle ACB = 37.6°, \angle ABC = 61.5°$
7 b $13\sqrt{3}$
8 b (1, 5, 2)
9 $-\frac{1}{2}, 3$
10 $(\frac{11}{9}, -\frac{26}{9}, -\frac{2}{9})$

Exam practice 6 (page 77)

1 a (2, 2, −2) b $\frac{1}{3}$
c $|\overrightarrow{AB}| = |\overrightarrow{BC}| = 3\sqrt{2}$ d $6\mathbf{i} - 2\mathbf{j} + 2\mathbf{k}$
2 b $2\sqrt{21}$ c 21.1°
3 $3\mathbf{i} + \mathbf{j} + 3\mathbf{k}$
4 a (2, 5, 4) c $\frac{3\sqrt{14}}{2}$
5 a $\mathbf{r} = 2\mathbf{i} + \mathbf{j} - 3\mathbf{k} + \lambda(\mathbf{i} - \mathbf{j} + \mathbf{k})$ c 84°
6 b $5\mathbf{i} - \mathbf{k}$ d 1.5 km
7 b $\mathbf{r} = 3\mathbf{i} - \mathbf{j} + 2\mathbf{k} + \lambda(-4\mathbf{i} + 2\mathbf{j} + 7\mathbf{k})$
c $11\mathbf{i} - 5\mathbf{j} - 12\mathbf{k}$
d 28.0°
8 a $\frac{4}{5}$ b $\frac{3}{2}$ c $(\frac{1}{2}, -\frac{1}{2}, -2)$
9 a 2 b $4\mathbf{i} + 7\mathbf{j} - 5\mathbf{k}$ d $12\mathbf{i} + 17\mathbf{j} + \mathbf{k}$ or $-4\mathbf{i} - 3\mathbf{j} - 11\mathbf{k}$
10 a $\mathbf{r} = 3\mathbf{i} + 5\mathbf{j} - \mathbf{k} + \lambda(-\mathbf{i} + 13\mathbf{j} - 13\mathbf{k})$
c $4\mathbf{i} - 8\mathbf{j} + 12\mathbf{k}$
d 116.9°
e 39.5
11 a (−4, −1, 5) b −2 c 82.6° (1 d.p.)
12 a $\mathbf{r} = 4\mathbf{j} - 2\mathbf{k} + \lambda(3\mathbf{i} - 6\mathbf{j} + 9\mathbf{k})$ b $\mathbf{i} + 2\mathbf{j} + \mathbf{k}$
c $3\mathbf{i} - 6\mathbf{j} + 9\mathbf{k}$ d $6\sqrt{21}$

Practice exam paper (page 80)

1 $-\frac{1}{2}x\cos 2x + \frac{1}{4}\sin 2x + C$
2 a $p = 1.910, q = 2.172$
b 3.85 (2 d.p.)
3 b 4 990 000
4 a $\frac{(8x^3 - 1)^5}{120} + C$
b $\frac{3\pi}{40}$
5 a $\frac{1}{4} + \frac{1}{4}x + \frac{3}{16}x^2 + \frac{1}{8}x^3$
b $1 + \frac{5}{4}x + x^2$
6 a $6\mathbf{i} - 6\mathbf{j} + 3\mathbf{k}$
b $\mathbf{r} = -2\mathbf{i} + 3\mathbf{j} - 4\mathbf{k} + \lambda(6\mathbf{i} - 6\mathbf{j} + 3\mathbf{k})$ or equivalent
d 35.7° (1 d.p.)
7 b $y - \sqrt{2} = -\frac{\sqrt{2}}{6}(x - 1)$ or equivalent
c 3
8 a $A = 1, B = 1, C = -2$
b $\frac{\sqrt{3}e^2}{4} \approx 3.2$

SINGLE USER LICENCE AGREEMENT FOR CORE 4 FOR EDEXCEL CD-ROM

IMPORTANT: READ CAREFULLY

WARNING: BY OPENING THE PACKAGE YOU AGREE TO BE BOUND BY THE TERMS OF THE LICENCE AGREEMENT BELOW.

This is a legally binding agreement between You (the user or purchaser) and Pearson Education Limited. By retaining this licence, any software media or accompanying written materials or carrying out any of the permitted activities You agree to be bound by the terms of the licence agreement below.

If You do not agree to these terms then promptly return the entire publication (this licence and all software, written materials, packaging and any other components received with it) with Your sales receipt to Your supplier for a full refund.

YOU ARE PERMITTED TO:

- Use (load into temporary memory or permanent storage) a single copy of the software on only one computer at a time. If this computer is linked to a network then the software may only be used in a manner such that it is not accessible to other machines on the network.
- Transfer the software from one computer to another provided that you only use it on one computer at a time.
- Print a single copy of any PDF file from the CD-ROM for the sole use of the user.

YOU MAY NOT:

- Rent or lease the software or any part of the publication.
- Copy any part of the documentation, except where specifically indicated otherwise.
- Make copies of the software, other than for backup purposes.
- Reverse engineer, decompile or disassemble the software.
- Use the software on more than one computer at a time.
- Install the software on any networked computer in a way that could allow access to it from more than one machine on the network.
- Use the software in any way not specified above without the prior written consent of Pearson Education Limited.
- Print off multiple copies of any PDF file.

ONE COPY ONLY

This licence is for a single user copy of the software

PEARSON EDUCATION LIMITED RESERVES THE RIGHT TO TERMINATE THIS LICENCE BY WRITTEN NOTICE AND TO TAKE ACTION TO RECOVER ANY DAMAGES SUFFERED BY PEARSON EDUCATION LIMITED IF YOU BREACH ANY PROVISION OF THIS AGREEMENT.

Pearson Education Limited and/or its licensors own the software.
You only own the disk on which the software is supplied.

Pearson Education Limited warrants that the diskette or CD-ROM on which the software is supplied is free from defects in materials and workmanship under normal use for ninety (90) days from the date You receive it. This warranty is limited to You and is not transferable. Pearson Education Limited does not warrant that the functions of the software meet Your requirements or that the media is compatible with any computer system on which it is used or that the operation of the software will be unlimited or error free.

You assume responsibility for selecting the software to achieve Your intended results and for the installation of, the use of and the results obtained from the software. The entire liability of Pearson Education Limited and its suppliers and your only remedy shall be replacement free of charge of the components that do not meet this warranty.

This limited warranty is void if any damage has resulted from accident, abuse, misapplication, service or modification by someone other than Pearson Education Limited. In no event shall Pearson Education Limited or its suppliers be liable for any damages whatsoever arising out of installation of the software, even if advised of the possibility of such damages. Pearson Education Limited will not be liable for any loss or damage of any nature suffered by any party as a result of reliance upon or reproduction of or any errors in the content of the publication.

Pearson Education Limited does not limit its liability for death or personal injury caused by its negligence.

This licence agreement shall be governed by and interpreted and construed in accordance with English law.